# Frontiers in South and Southeast Asian Development Research

This book series publishes monographs and edited volumes devoted to studies on various aspects of development and policies in South and Southeast Asia. The books cover in-depth analyses of individual countries, regions, sectors, and comparative cross-country studies. They include both a specific and a general focus on the latest advances of the various aspects of development. In cross-country studies, the geographical scope will be inclusive to the developed Asian economies when it comes to issues like security, economic cooperation, energy, finances, trade, environment, and technology transfer. The books in this series are expected to go beyond the traditional concepts of inequality, poverty, growth, and development and offer a broad, in-depth analysis and inclusive view on sustainable development and well-being. The series provides a platform for researchers to carry out rigorous analyses, to promote, share, and discuss issues, findings, and perspectives in various areas. The topics may include, but not limited to agriculture, industry, services, business, governance, institutions, politics, security, globalization, international relations, gender, culture, religion and society, economics, and social development strategies of South and Southeast Asia. Not only that, given the current pandemic and its multifaceted impacts on trade, production, supply chain and organization of economic activities, this series also purports to highlight changes in research on development economics in the post-pandemic timeframe. Frontiers in South and Southeast Asian Development Research allows for a deeper appreciation of the various issues around Asian development with high quality and peer reviewed contributions. Volumes published in the series are important reading offering original research supported empirically which can serve as reference sources for researchers, educators, and graduate students, as well as consultants, business professionals, and policy makers, interested in the private and public sectors development of the Asian continent and its challenges and strategies for sustainable development.

Hanh-Thi Vu • Léo-Paul Dana
Editors

# Vietnam on the Road to Sustainable Economic, Social and Business Development

Springer

*Editors*
Hanh-Thi Vu
Center for Foreign Economic Research, Training and Consultancy
Foreign Trade University
Hanoi, Vietnam

Léo-Paul Dana
Faculty of Management
Dalhousie University
Halifax, NS, Canada

ISSN 2948-1279 ISSN 2948-1287 (electronic)
Frontiers in South and Southeast Asian Development Research
ISBN 978-981-96-4370-7 ISBN 978-981-96-4371-4 (eBook)
https://doi.org/10.1007/978-981-96-4371-4



This Springer imprint is published by the registered company Springer Nature Singapore Pte Ltd.
The registered company address is: 152 Beach Road, #21-01/04 Gateway East, Singapore 189721, Singapore

# Acknowledgment

This book is sponsored by Foreign Trade University under research program number FTURP01-2020-06.

# Contents

# List of Figures

# List of Tables

# Part I
# Environmental Sustainability

# Vietnam 30 Years After 1995 Normalization

**Léo-Paul Dana**

**Abstract** Until 1975, war delayed business development in Vietnam, and from 1975 to 1994 a trade embargo further limited development. In 1995, normalization of relations with the United States opened many doors. Vietnam, today, is a rapidly growing nation on the road to sustainable economic, social, and business development.

## 1 Introduction

My first visit to Vietnam was in 1992, scouting for manufacturers on behalf of a firm in Canada. As wages were rising in Canada, many Canadian producers were looking to outsource. Other than the very young, few people in Vietnam spoke English, but elderly people who had studied under French colonial rule spoke eloquent French with me, and younger people who had studied in East Germany conversed with me in German.

The Vietnamese work ethic was evident (Exhibit 1), but much business was informal (Exhibit 2) and infrastructure was poor. I wrote, "Unlike the former Soviet Union and Eastern Europe, Vietnam has chosen to retain socialist ideals, but the country appears to be thriving more on free enterprise than on Marxist ideology. There is a constant buzz of mercantile energy. Entrepreneurs optimise the use of their minimal resources, and carry on despite a poor infrastructure (Dana, 1994b, p. 23)."

Postage stamps did not have adhesive, so people spent much time brushing paste on stamps to affix them to envelopes (Exhibit 3). Credit cards were rare, and people walked around with low-denomination banknotes (Exhibit 4). There was not much motor vehicle traffic in Vietnam; vendors were often ambulant (Exhibit 5). Few

L.-P. Dana (✉)
Faculty of Management, Dalhousie University, Halifax, NS, Canada
e-mail: lp762359@dal.ca

 
H.-T. Vu, L.-P. Dana (eds.), *Vietnam on the Road to Sustainable Economic, Social and Business Development*, Frontiers in South and Southeast Asian Development Research, https://doi.org/10.1007/978-981-96-4371-4_1

**Exhibit 1** Working hard with Singer sewing machines; photograph by Léo-Paul Dana

homes had refrigerators, so ice was delivered to families (Exhibit 6); a popular means of transportation for people and merchandise was the cyclo (Exhibit 7)—a 1939 invention created by a Frenchman. Remnants of French colonialism, many vehicles were French (Exhibit 8). The fleet of Vietnam Airlines was entirely from the Soviet Union (Exhibit 9) and comfort was limited (Exhibit 10). Flights were few and trains were crowded (Exhibit 11).

I observed: "In contrast to reform in Eastern Europe and the former Soviet Union, where a market economy was decreed to have replaced communism, liberalization in Vietnam has involved ushering in entrepreneurship as a complement to state enterprise, rather than as a replacement of Marxist ideals. Without overthrowing the socialist establishment, doi moi (literally 'renovation'), the Vietnamese version of perestroika is allowing small businesses to play an increasingly important role in the economic development of the nation (Dana, 1994a, p. 95)."

## 1.1 Toward a New Era

In 1994, the lifting of the U.S. embargo paved the way for a new Vietnam. In February 1994, Coca-Cola returned to Vietnam. The following year, Nike began manufacturing in Vietnam and the United States normalized relations with Vietnam. Reebok also began producing in Vietnam (Exhibit 12).

**Exhibit 2** Informal vending; photograph by Léo-Paul Dana

In 1997, with Professor Richard W. Wright, I took a class of undergraduate and MBA students from McGill University on a study tour of Vietnam; this was the first time McGill offered Faculty of Management courses in Asia. Shortly thereafter, I served as Visiting Professor on the University of Pittsburgh Semester at Sea program, sailing with several hundred North American students to Vietnam.

In 1998, Vietnam's Central Committee convened to outline a new strategy. In contrast to policy that had previously designated entrepreneurs as economic engines of growth, the Central Committee decided that state-owned enterprises would lead the Vietnamese economy (Dana, 2002, p. 242).

I observed that the country was on a harmonious route to development and wrote, "This concept of harmony—between government firms still operating under a system of central planning, and private enterprise being a complement to the state sector—is also characteristic of Vietnam's Doi-Moi Model. The Vietnamese word doi-moi literally means "renovation." No major political reform is implied here; instead, this model is compatible with restoring the prestige of the ruling Communist Party of Vietnam. The government affirmed its commitment to free enterprise within the context of socialism (2002, p. 8.)."

The new millennium witnessed rapid development in Vietnam. Ciputra Hanoi was built to house up to 50,000 people in a green eco-urban area (Exhibit 13). New wealth was reflected in the new 500,000-dong banknotes (Exhibit 14) in circulation

**Exhibit 3** Pasting postage stamps onto envelopes; photograph by Léo-Paul Dana

since 2003. I wrote, "Vietnam's model of economic transition has been stunning (2014, p. 338)."

Towering high above the Tran Phu Building (Exhibit 15) and the Cầu Giấy District of Hanoi (Exhibit 16), the D'Eldorado housing project (Exhibit 17) was completed in 2019. The FPT Telecom Tower (Exhibit 18) was completed in 2021 and the Vinacomin Tower (Exhibit 19) was completed in 2024. Symbolic of new prosperity, in 2024 a new office complex of the District Party Committee (Exhibit 20) was inaugurated in Hanoi; at a cost exceeding 434 billion dong, this includes 185 functional rooms.

The year 2024 marked the 70th anniversary of the withdrawal of French troops (Exhibit 21). Some billboards featured Communist revolutionary Ho Chi Minh as national hero (Exhibit 22).

## *1.2 Vietnam Today*

The Temple of Literature (Exhibit 23), established in 1070, hosted the first national university of Vietnam—the Imperial Academy—from 1076 to 1779 and Vietnamese people continue to value education (Exhibit 24). On the occasion of their graduation, many students come to the Imperial Citadel of Thang Long (Exhibit 25). A free trade agreement was signed in 2023, between Vietnam and Israel, also a country with a strong focus on education and innovation, coupled with a robust startup

**Exhibit 4** Low-denomination banknotes; photograph by Léo-Paul Dana

culture that is supported by a strong venture capital industry; cooperation between these two countries is very promising.

Urban areas of Vietnam today have an international and modern feel (Exhibit 26). Some remnants of the French Empire remain (Exhibit 27), but culturally Vietnam identifies with its rich heritage (Exhibit 28). Traditional Vietnamese cuisine is as popular as ever (Exhibit 29).

Vietnam is thriving on free enterprise rather than on Marxist ideology. The hammer and sickle can be seen across Vietnam (Exhibit 30), but unlike countries of Eastern Europe that consider the hammer and sickle as the symbol of a communism under totalitarian ideology, in Vietnam the hammer and sickle symbol is symbolic of the unity between industrial workers and farmers.

From an economic and business perspective, it is evident that Vietnam is on the road to sustainable development. Founded in 2017 by entrepreneur Pham Nhat Vuong to manufacture electric vehicles in Vietnam, in 2021 VinFast became the first Vietnamese automaker to export cars (Exhibit 31) to the United States.

**Exhibit 5** Ambulant vendors; photograph by Léo-Paul Dana

**Exhibit 6** Ice delivery; photograph by Léo-Paul Dana

**Exhibit 7** Driver seated behind passenger on cyclo; photograph by Léo-Paul Dana

**Exhibit 8** A Peugeot; photograph by Léo-Paul Dana

**Exhibit 9** Yakovlev Yak-40; photograph by Léo-Paul Dana

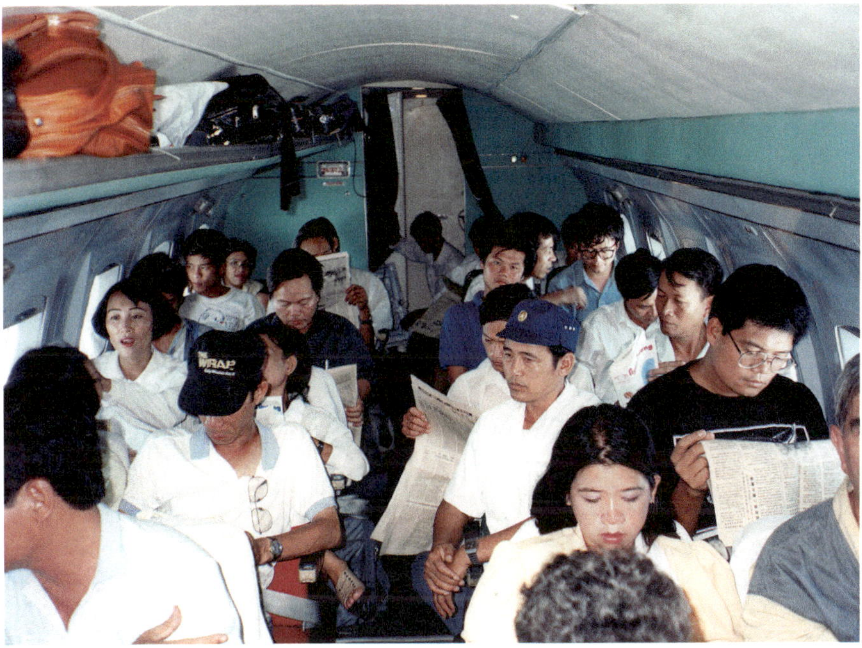

**Exhibit 10** Aboard the Yak-40; photograph by Léo-Paul Dana

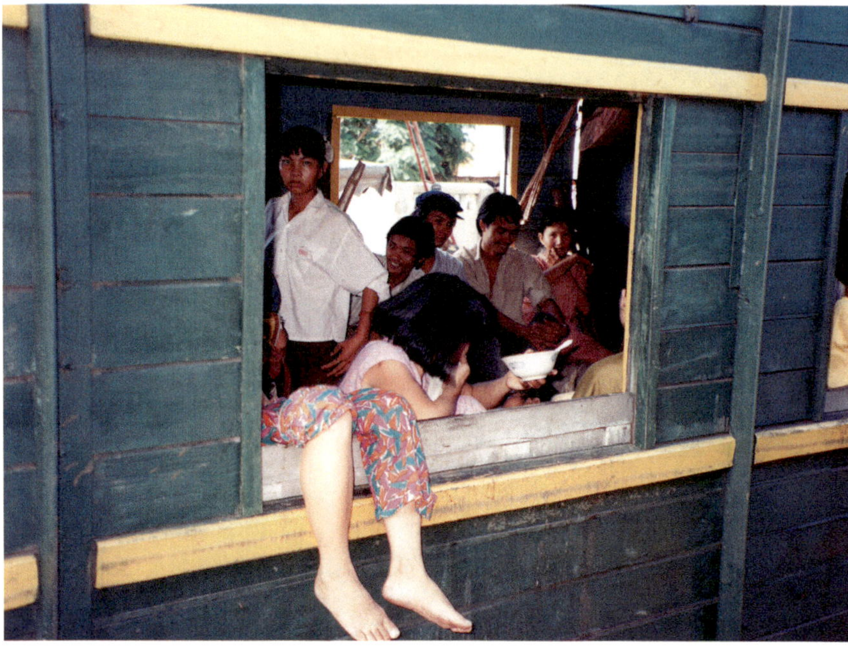

**Exhibit 11** Crowded train; photograph by Léo-Paul Dana

**Exhibit 12** Manufacturing Reeboks in Vietnam; photograph by Léo-Paul Dana

**Exhibit 13** Ciputra Hanoi; photograph by Léo-Paul Dana

**Exhibit 14** 500,000-dong bank notes; photograph by Léo-Paul Dana

**Exhibit 15** Tran Phu Building; photograph by Léo-Paul Dana

**Exhibit 16** Cầu Giấy District of Hanoi; photograph by Léo-Paul Dana

**Exhibit 17** D'Eldorado residential building; photograph by Léo-Paul Dana

**Exhibit 18** FPT Telecom Tower; photograph by Léo-Paul Dana

**Exhibit 19** Vinacomin Tower; photograph by Léo-Paul Dana

**Exhibit 20** New offices of the District Party Committee in Hanoi; photograph by Léo-Paul Dana

**Exhibit 21** The year 2024 marked 70 years of liberation from France; photograph by Léo-Paul Dana

**Exhibit 22** Ho Chi Minh as national hero; photograph by Léo-Paul Dana

**Exhibit 23** Temple of Literature; photograph by Léo-Paul Dana

**Exhibit 24** Graduating class; photograph by Léo-Paul Dana

**Exhibit 25** The Imperial Citadel of Thang Long; photograph by Léo-Paul Dana

**Exhibit 26** Hanoi; photograph by Léo-Paul Dana

**Exhibit 27** *Maison du Jade*, dating from 1926; photograph by Léo-Paul Dana

**Exhibit 28** Wearing traditional clothes for performance of traditional music; photograph by Léo-Paul Dana

**Exhibit 29** Restaurant serving traditional Vietnamese cuisine; photograph by Léo-Paul Dana

**Exhibit 30** Hammer and sickle; photograph by Léo-Paul Dana

**Exhibit 31** Made in Vietnam; photograph by Léo-Paul Dana

## References

Dana L-P (1994a) A marxist mini-dragon? Entrepreneurship in today's Vietnam. J Small Bus Manag XXXII(2):95–102

Dana L-P (1994b) Economic reform in the new Vietnam. Curr Aff LXX(11):19–25

Dana L-P (2002) When economies change paths: models of transition in China, the Central Asian Republics, Myanmar, and the Nations of Former Indochine Française. World Scientific, Singapore, London and Hong Kong

Dana L-P (2014) Asian models of entrepreneurship – from the Indian Union and Nepal to the Japanese Archipelago: context, policy and practice, 2nd edn. World Scientific, Singapore and London

**Léo-Paul Dana** is a Professor at Dalhousie University. A graduate of McGill University, he has served as Marie Curie Fellow at Princeton University and Visiting Professor at INSEAD and at Kingston University.

# Sustainable Energy for Future Development in Vietnam

**Hanh-Thi Vu and Ha-Ngoc Van Nguyen**

**Abstract** This chapter discusses several issues of the energy sector in Vietnam, including energy structure, fossil fuel, and renewable energy for sustainable development. The demand for energy is growing rapidly in recent years and will continue to increase until 2050. The importance of energy security to socialeconomic development and the requirement of environmental protection are the key factors of the development of renewable and sustainable energy sources.

## 1 Introduction

Over the years, Vietnam has made great efforts to implement and promote sustainable development goals to be integrated into economic development programs and has achieved many results, which are highly appreciated by the international community. Vietnam has always committed to implementation goals under the 2030 Agenda for Sustainable Development of the United Nations. In 2020, facing unprecedented challenges in history such as the Covid-19 pandemic, natural disasters, and floods in the central region, Vietnam increasingly recognized the importance of sustainable development, especially the sustainable development of energy in the future. The journey to achieve the desired results requires necessary adjustments, but the ultimate goal still remains: "no one is left behind." Together, we strive to create a better and cleaner life for everyone today and for future generations.

H.-T. Vu
Center for Foreign Economic Research Training and Consultancy, Foreign Trade University, Dong Da, Ha Noi, Vietnam
e-mail: hanhvt@ftu.edu.vn

H.-N. Van Nguyen (✉)
GIZ Energy Support Programme, Tay Ho, Ha Noi, Vietnam
e-mail: ha.nguyenngoc@giz.de



H.-T. Vu, L.-P. Dana (eds.), *Vietnam on the Road to Sustainable Economic, Social and Business Development*, Frontiers in South and Southeast Asian Development Research, https://doi.org/10.1007/978-981-96-4371-4_2

In the energy sector, the Politburo has issued a number of important resolutions, including Resolution No. 18-NQ/TW, dated October 25, 2007 of the 10th Politburo on "Strategic Orientation for Vietnam's national energy development to 2020, vision to 2050" and Conclusion No. 26-KL/TW, dated October 24, 2003 of the 9th Politburo on "Strategy and planning for developing Vietnam electricity." After more than 10 years of implementing Resolution 18 and 15 years of implementing Conclusion No. 26, national energy development requires new guidelines and policies to match development trends and requirements of the socioeconomic 10-year period 2021–2030, so the Politburo assigned the Central Committee of Economic Affairs to assume the prime responsibility and coordinate with agencies in formulating a Project on summarizing the above resolutions and conclusions. On the basis of this project, on February 11, 2020, the Politburo issued Resolution No. 55-NQ/TW on "Strategic orientations for Vietnam's national energy development to 2030, with a vision to year 2045." In particular, the Resolution clearly identifies the main strategy in the coming time, which is to focus on developing diverse and suitable energy sources, while giving priority to renewable energy. Resolution 55 introduces several new features aimed at directing national energy development, ensuring national energy security, stimulating the process of industrialization and modernization of the country, meeting requirements for fast and sustainable economic development, promoting independence and self-reliance associated with international integration, ensuring national defense and security, protecting the ecological environment and enhancing access to energy at reasonable cost and service.

Sustainable energy development in response to climate change and green growth orientation: GHG emission estimates for 2020 and 2030 show that GHG in the energy sector is still the main source of emissions with 381.1 million tons of $CO_2$ equivalent in 2020 and 648.5 million tons in 2030. Realizing the significant growth of national greenhouse gas emissions in the future and fulfilling national responsibility, in the process of mitigating the impacts of global warming, Vietnam has set GHG reduction targets in its Green Growth Strategy, Nationally Determined Contribution (NDC), and Vietnam Renewable Energy Development Strategy.

## 2 Energy Security

Ensuring the world's energy supply is increasingly challenging, so the topic of energy security has become one of the top priorities of countries around the world. Along with the momentum of social development, the demand for energy use for economic development tends to increase. According to the forecast of the Ministry of Industry and Trade, the electricity demand for economic development from 2021 onward will still grow at a high rate of 8–10%/year, while fossil energy sources such as oil, gas, and coal are draining away. Therefore, using energy economically, efficiently, and finding new energy sources in the future are effective and sustainable solutions.

Previously, thermoelectricity was one of the solutions to this problem. With the advantage of cheaper production costs compared to other types of renewable energy, thermal power has been included in the Power Plan VII by the management with a large proportion. However, after identifying the drawbacks that coal-fired power plants have caused to the environment, coal power is no longer a top priority in the current development strategy.

According to a report by the Department of Petroleum and Coal (Ministry of Industry and Trade), it has become more difficult to find and exploit three types of fossil raw materials, namely, crude oil, natural gas, and coal, and Vietnam has been increasingly importing these materials in recent years. For example, the exploitation output of crude oil had reached 15.2 million tons in 2016; however, it dropped to only 8.1 million tons in 2024. Gas production also fell rapidly, from 10.61 billion m3 in 2016 to 6.32 billion m3 in 2024 (Duy, 2020; Huong, 2025). Meanwhile, coal imports of Vietnam increased sharply in the period from 2015–2024 from 7 million tons to 63.8 million tons (Duy, 2020; Huong, 2025). The lack of energy sources is the problem not only for Vietnam but also for many countries around the world in ensuring energy security for production and consumption.

With a considerable need for investment in the clean energy sector, Resolution 55 clearly states that it is necessary to develop breakthrough mechanisms and policies to encourage and promote the strong development of renewable energy sources, targeting to maximize the replacement of fossil energy sources with renewable energy ones. This process of institutionalization requires legal documents, mechanisms and policies to encourage investment, electricity pricing policies, participation in investment in the power transmission system, policies to encourage households to use electricity economically and adoption of clean and renewable energy, especially in industry and transportation, as well as the use of electricity in line with the general trend in the world. In addition, it is necessary to improve the legal basis to effectively form and operate funds for sustainable energy development, promoting economical and efficient use of energy in the direction of socialization. In the long term, research is needed to promulgate a separate law on renewable energy.

While Resolution 18 mentioned the point of view "Promote the elimination of subsidies and monopolies, toward the complete abolition of the implementation of social policies through energy prices," Resolution 55 stated clearly: "To encourage and create all favorable conditions for all economic sectors, especially the private sector, to participate in energy development; resolutely eliminate all manifestations of subsidies, monopolies, unequal competition, and lack of transparency in the energy industry." This is a very new point of view, which is considered to be suitable with the actual situation and accurately reflects the role of the private sector in economic development.

In the near future, to encourage the involvement of domestic and foreign private enterprises in energy development investment, first of all, stronger and more comprehensive support policies are necessary to accelerate the market implementation roadmap. In a competitive electricity market, there should be a mechanism for direct electricity purchase and sale agreements between the two parties, namely the producer and the consumer, a mechanism for bidding and auction of energy supply, as well as transparent electricity pricing. Besides, it is necessary to encourage and attract energy investment projects in the form of public-private partnership (PPP); implement a flexible and efficient credit policy to facilitate energy enterprises to access capital sources, prioritize enterprise project green energy, and create optimal conditions for energy development.

It can be affirmed that renewable energy is said to be the solution to the problem of national energy security in the future.

## 3 Fossil Fuel and Pollution

Fossil fuels are fuels formed by the anaerobic decomposition of dead organisms buried more than 300 million years ago. These feedstocks contain high levels of carbon and hydrocarbons and exist in the form of coal, petroleum, and natural gas.

According to the Power Development Plan VIII, by 2030, fossil fuel still remains the main sources of power. Among the total installed capacity of 150,489 MW, coal thermal power contributes the largest proportion of 20%, followed by hydroelectricity with 19.5%, while LNG and domestic gas thermal power accounts for 14.9% and 9.9%, respectively. However, due to its undeniable negative impacts on the environment, especially in recent years, fossil energy is no longer encouraged by the Government, which reflects a major change compared to the Power Development Plan VII, and Vietnam is also shifting towards other sustainable energy sources.

Coal is a very important fuel in daily life and production. In 2024, 63.8 million tons of coal was imported, twice as much as in 2022, and most of which was for electricity generation (Huong, 2025). As in the Decision 839/QD-TTg 2023, from 2021 to 2030, it is forecasted that coal imports would keep increasing, reaching 73 million tons. The demand for imported coal of thermal power plants would rise sharply until 2035, before starting to drop in the following phase (Fig. 1).

Total coal production forecasted in the 2016–2025 period is 485.9 million tons. The total imported coal volume is 344 million tons. The total output of coal exploitation and import in the period 2016–2035 is 1,056.4 and 1,085.1 million tons, respectively. The optimal calculation results show that the domestic coal exploitation volume is fully mobilized according to the PDP with the exploited capacity reaching 53.8 million tons in 2025 and increasing to 60 million tons in 2035. Coal import volume will increase to 69.2 million tons in 2025 and 157.1 million tons in 2035.

Until 2025, electricity production will still account for the largest part of coal consumption with 475.3 million tons in the period 2016–2025 and will increase to 1,046.7 million tons in the period from 2026 to 2035 (Table 1).

The period 2011–2020 marks the growth of coal-fired power with an average rate of 18%/year. In 2010, coal-fired power accounted for 19% of the total installed capacity of power sources, equivalent to 3,941 GW, but by 2020 it had increased to more than 30%, equivalent to 20,867 GW, which implied a strong growth of

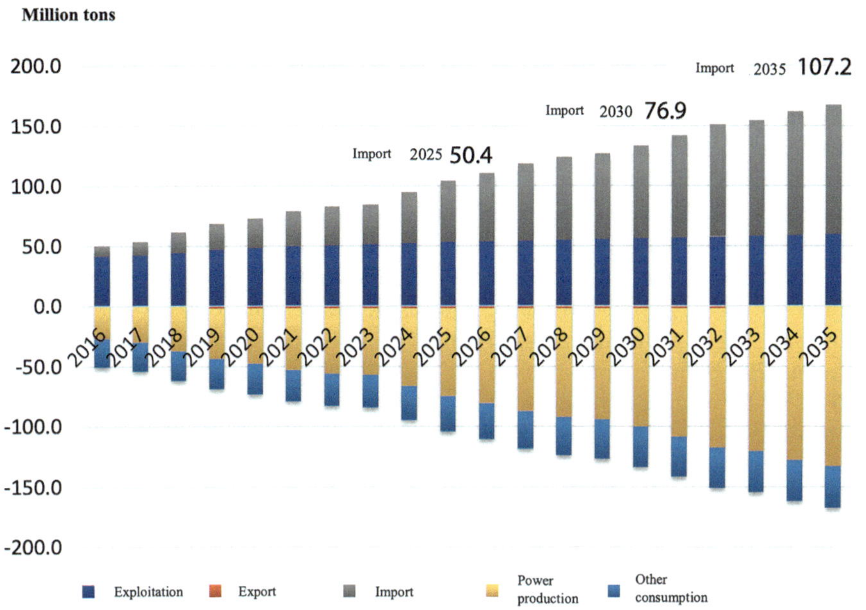


**Fig. 1** Coal supply-demand balance in the proposed scenario. Source: Danish Energy Agency (2017)

**Table 1** Coal supply-demand balance in the period 2016–2035 (million tons)

| | 2016 – 2025 | 2026–2035 | 2016–2035 |
|---|---|---|---|
| Exploitation | 485.9 | 570.5 | 1,056.4 |
| Export | 14.8 | 14.3 | 29.1 |
| Import | 266.2 | 818.9 | 1,085.1 |
| Electricity production | 475.3 | 1,046.7 | 1,522.0 |
| Others | 262.1 | 328.3 | 590.4 |

Source: Danish Energy Agency (2017)

coal-fired power. The growth rate of coal-fired power plants from 2010 to 2015 was 27% and from 2016 to 2020 was 10%/year. The reason for this decline was that coal-fired power plants faced many difficulties during construction. The factories using imported coal are mainly located in the south. The northern region concentrated on thermal power plants using domestically exploited materials with a total capacity of more than 12 GW; however, many plants having been operating for many years with poor performance. The average mobilization level is about 6,000h/year for coal-fired power plants, which is considered relatively high by experts and this level tends to increase gradually in the following years. This proves that the power system begins to lack backup power. In the next year, it is necessary to enhance the maintenance of thermal power plants to ensure that they meet the needs outlined.

Besides coal, natural gas is also a vitally important source of fossil fuels, with a much lower level of emission produced compared to others such as oil or coal. Thus, the current demand for liquefied natural gas (LNG) in our country is always high. With the option of domestic natural gas exploitation and the shortfall being compensated by imported LNG, imported LNG output is expected to increase rapidly in the period after 2025. Imported LNG output will increase to 1.1 million tons with the operation of Thi Vai LNG port in 2025 and 11.7 million tons by 2035. The balance of supply and demand of natural gas is shown in Fig. 2 and Table 2:

Total domestic gas exploitation volume in the 2016–2025 period reaches 125.2 billion $m^3$. In the period 2016–2035, gas exploitation volume is 270.6 billion $m^3$ (covering 92% of the total mobilized natural gas in the baseline option of gas supply of about 295 billion, approximately 295 billion $m^3$). Gas supply volume is 109.3 billion $m^3$ for electricity production and 23.7 billion $m^3$ for other consumers in the 2016–2025 period. This volume would increase to 200.4 billion $m^3$ for the production of electricity in 2035.

In terms of gas-fired power plants: most of them are combined-cycle gas turbine plants (CCGT) operating with a life cycle of 10–18 years, generally having advanced technology equipment (gas turbines of generation E, F of Mitsubishi, GE, Siemens, Alstom) with high efficiency of 48–53% (in particular, that of Ba Ria CCGT plant is 41%). Besides, all factories commonly use modern distributed control systems (DCS) for the main equipment systems. SOx and dust emissions at gas-fired thermal

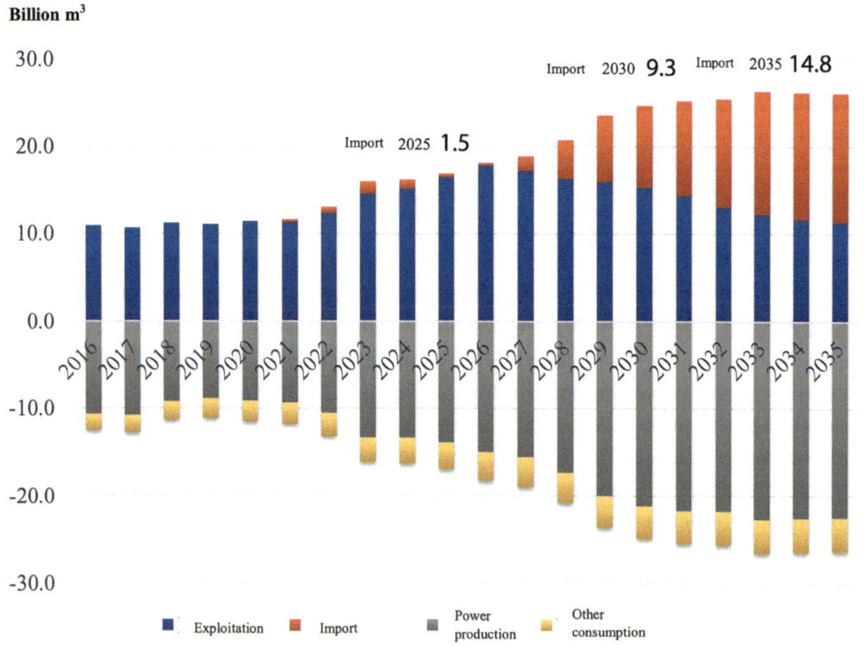


**Fig. 2** Gas supply-demand balance in the period 2016–2035. Source: Danish Energy Agency (2017)

**Table 2** Gas supply-demand balance in the periods in the proposed scenario (billion m$^3$)

| | 2016–2025 | 2026–2035 | 2016–2035 |
|---|---|---|---|
| Exploitation | 125.2 | 145.4 | 270.6 |
| Import | 4.2 | 90.2 | 94.4 |
| Electricity production | 109.3 | 200.4 | 309.7 |
| Other consumers | 23.7 | 36.2 | 60.0 |

Source: Danish Energy Agency (2017)

**Table 3** Crude oil supply-demand balance in the proposed scenario (million tons)

| | 2016–2025 | 2026–2035 | 2016–2035 |
|---|---|---|---|
| Exploitation | 136.5 | 42.6 | 179.1 |
| Import | 103.4 | 330.6 | 434.0 |
| Export | 51.6 | – | 51.6 |
| Oil refineries | 185.4 | 373.2 | 558.6 |

Source: Danish Energy Agency (2017)

power plants are nonexistent, apart from NOx emissions that have been eliminated at plants at Phu My Power Center, but not been treated at Nhon Trach 1 & 2 plants, Ba Ria and Ca Mau. CCGT plants can use more oil fuel in case of gas shortages. Each CCGT plant has a combination of 2 or 3 gas turbine units and 1 steam turbine unit with a small capacity (<300 MW), so it can operate well at low capacity (about 20% of total capacity).

Besides coal and gas, oil is also an important source of fossil fuels, being a highly volatile commodity in the world market. The extracted crude oil is fully mobilized as in the proposed plan for domestic and foreign crude oil extraction. The output of crude oil imports depends on the progress of construction and operation of oil refineries in the planning stage. According to the calculation results of the optimal energy system, crude oil demand of refineries is as follows: for 2 Dung Quat oil refineries (including expansion projects) and Nghi Son refinery, the crude oil amount needed for oil refining is about 18.2 million tons/year. Crude oil demand will increase to 28.0 million tons with the operation of Vung Ro oil refinery by 2025, and to 41.5 million tons/year when Nam Van Phong refinery comes into operation by 2030. With the current assumptions about energy prices and investment costs, Nghi Son and Long Son oil refineries will not operate in the period to 2035. The crude oil supply-demand balance is shown below here (Table 3):

The total output of crude oil exploited reaches 136.5 million tons in the period 2016–2025. In the whole period 2016–2035, crude oil production is 179.1 million tons m$^3$ (reaching 100% of crude oil mobilized according to the plan for crude oil supply). Crude oil supply volume for oil refineries is 185.4 million tons. Crude oil exports will gradually decrease until 2024 to provide fuel for domestic refineries. In 2025, in the context of low demand from oil refineries, crude oil export will reach 51.6 million tons, generating foreign currency revenues for Vietnam. With the operation of the refinery in Nghi Son using imported crude oil, crude oil import is expected to rise to 18 million tons in 2025 and 38 million tons in 2035. The proposed scenario uses the option of using domestic and foreign crude oil exploitation as in the Petroleum Planning. Thus, should the required level of crude oil exploitation not be achieved, the demand of the refinery for imported crude oil will be higher (Danish Energy Agency, 2017).

Currently, thermal power plants, oil-fired thermal plants, and electrical system still have to rely on fuel oil for thermal power gas turbines and diesel oil in the gas phase breakdowns, maintenance of PM3 CAA air system, and gas system maintenance. At the same time, fuel oil thermal power and diesel oil gas turbines are also mobilized in the dry season months (March, April, and May). Due to the low number of hours mobilized, the downtime rate due to machine breakdown and maintenance is also low. Despite the longtime of operation (20–40 years), oil-fired thermal power plants such as Thu Duc, Can Tho, and Hiep Phuoc plants still have to continue in the near future to reduce the risk of power shortage. Oil-fired thermal power plants are mostly old plants, so the emission amount of SOx, NOx, and PM are quite high (10 times higher than the coal-fired power plants). These are sources of greenhouse gas that pollute the environment.

Gas-fired and coal-fired thermal power considerable change in the proportion of installed capacity of the power supply structure. In 2010, the installed capacity was 7,846 GW, equivalent to 38%, however, by 2023, it dropped sharply to 10.3% with a capacity of 8,297 GW (EVN Department of Communication, 2024). Gas and oil thermal power plants are concentrated in the South with about 7 GW of combined cycle gas turbine plants.

It can be seen that today, in real-world production and consumption of energy in Vietnam, the objective of ensuring national energy faces many challenges. Sources of domestic supply are insufficient to meet the demand for manufacturing and consumption, leading to an imbalance in the structure of the energy industry when energy imports are increasing. Some indicators of energy security are fluctuating in an unfavorable direction. For example, the falling amount of coal and oil production reserves; the increasing dependence on imports of coal, oil, and natural gas; and the increasing share of coal, oil, and natural gas import costs in total domestic income.

The current policy is gradually shifting to green, clean, and renewable energy instead of prioritizing development of fossil energy (such as coal, oil, and gas) as before. It is expected that the proportion of coal-fired power in the industry structure will decrease from 34% in 2020 to 20% in 2030. Solar power's proportion is estimated to drop to 8.5% by 2030, while wind power would increase to 14.5% due to higher efficiency. Besides, the use of LNG would also decrease to 14.9% as a result of increasing import price (PDP VIII).

Sustainable energy development in response to climate change and green growth orientation is critically important and the results of GHG emission estimates in 2020 and 2030 show total GHG emissions in four sectors (energy, agriculture, LULUCF, and waste) in 2010 were 225.6 million tons of $CO_2$ equivalent, increasing to 466 million tons in 2020, and about 760.5 million tons in 2030. The energy sector remains the largest GHG emission source with 381.1 million tons of $CO_2$ equivalent in 2020 and 648.5 million tons in 2030 (PDP VIII), in which, electricity production using fossil fuels is a main contributor. Thus, the emerging issue is to transition to green and environmentally friendly energy to reduce fossil fuel consumption, mitigating the impact of global warming phenomenon. Fulfilling national responsibility in environmental issues, Vietnam's government has included GHG reduction targets in its Green Growth Strategy, Development Strategy, and especially in its Nationally Determined Contribution Report (NDC Vietnam, updated 2020).

According to Vietnam's Nationally Determined Contribution (2020 update), with domestic resources by 2030, Vietnam will reduce total GHG emissions by approximately 9% compared to the scenario of business as usual (BAU), equivalent to 83.9 million tons of $CO_2$ equivalent. In which, the emissions from the energy sector are expected to fall to 51.5 million hundredweight of $CO_2$ equivalent, around 5.5% compared to the BAU scenario. However, the abovementioned contribution of 9% can be increased to up to 27% compared to the BAU scenario (about 250.8 million hundredweight of $CO_2$ equivalent) with international support received through bilateral, multilateral cooperation as well as through the implementation of new mechanisms under the Paris Agreement. In particular, the expected GHG decrease in the energy sector would be 155.8 million hundredweight of $CO_2$ equivalent, approximately 16.7% compared to the BAU scenario.

Vietnam's economy was one of the first to successfully recover during the period of controlling the Covid-19 pandemic, providing a solid foundation for the country to transition toward a more flexible, green, and sustainable economy. One of the tools the government is considering is carbon tax. Currently, market studies show that a carbon tax can be applied to high-emission sectors such as the energy sector, cement production, transportation, etc. In short, a carbon tax is expressed as a value such as VND for each ton of carbon released into the atmosphere, which is one of the tools to create economic development incentives for emitters to reduce their emissions into the environment without imposing rigid technical standards (Phan, 2024). Emissions are proportional to the amount to be paid, which is considered fair and enhances our country's energy security. The revenue from the carbon tax is expected to lead to increased spending and investment in low-carbon infrastructure, spurring the development of greener and cleaner technologies that deliver higher yields than coal-fired power such as renewable energy and nuclear energy.

## 4 Renewable and Sustainable Energy

Strategic documents on environmental protection and sustainable development goals have been integrated into the master plan, which is reflected in specific objectives: ensuring energy security while reducing the proportion of coal-fired power sources, developing power generation from nuclear energy to reduce fossil fuel consumption and GHG emissions, developing sustainably and environmentally friendly, and encouraging the development of renewable energy.

Renewable energy is useful energy derived from renewable natural resources that are replenished naturally over the cycle of human time, including carbon-neutral sources such as sunlight, wind, rain, tides, waves, and geothermal energy. Most types of renewable energy are sustainable except biomass energy.

Due to the impact of the incentive mechanism for solar power development in Decision No.13/2020/QD-TTg dated 6 April, 2020 and the incentive mechanism for wind power development in Decision No.39/2018/QD-TTg dated 9 October, 2018, wind and solar projects have grown significantly, resulting in a rapid increase in renewable energy capacity in operation. In the period from 2018–2019, solar power was rapidly developed, and by the end of 2020, total installed capacity of solar power reached 17 GW. Until 2023, renewable energy sources (including wind power and solar power) accounted for 26.9% of the total capacity (EVN Department of Communication, 2024), while oil and gas thermal power barely had any new development during the period from 2011–2020. However, the rapid growth of solar power projects exceeded the planning as in PDP VII, especially in the Central region and the Central Highlands area, leads to the unbalance in power proportion by sources and regions, laying a burden on the electrical grid (Government Inspectorate, 2023).

## 4.1 Solar Power

Talking about renewable energy, the first thing to mention is solar energy. Solar energy is photovoltaic and thermal energy coming from the sun that is harnessed and used through a series of constantly evolving technologies to convert energy from the sun into other forms such as electricity and heat. This is considered a clean energy source, and can be an alternative source for fossil fuels, such as oil and gas. Current research and seminars have provided many optimal solutions to handle faulty or damaged solar panels so that they do not affect the environment.

Currently, three technologies have been developed and commercialized including solar photovoltaic (SPV), concentrated solar power (CSP), and solar thermal technology (STT). The most important use of solar energy today and in the future is still electricity generation. In Vietnam, two types of solar power generation technologies are widely developed, namely, CSP and SPV. In particular, SPV includes crystalline solar cells (90% market share) and thin film solar cells (about 10% market share). CSP uses a system of multiple glass panels, reflectors, and tracking systems to focus sunlight from a large area into a small area.

Currently, solar power technology has been developed for a relatively longtime, so the investment cost for construction and installation of solar panels has been optimized. Thus, the price is much cheaper while the energy conversion efficiency is increasing (efficiency of panels reaches 30–35% in actual operating conditions in 2030). The solar modules currently on the market are about 15–17% efficient, producing about 150–170 Wp DC per $m^2$ with Wp as the unit of measurement of the solar energy generated. Besides, the construction time of solar power projects is quite short, only about 6–12 months, leading to several households' investment in solar power for sale to EVN.

In 2020, our country's solar power has observed many new developments. Decision 13/2020/QD-TTg of the Prime Minister stipulated that the purchase price of rooftop solar power was higher than that of other types of solar power, showing the Government's interest and focused development in rooftop solar power. However, the government's Decision 13 on the incentive mechanism for solar power in Vietnam was only applied for system whose power generate and electricity meter reading confirmation dates from July 1, 2019 to December 12, 2020. Investors and firms had rushed to connect to the power grid to take advantage of the preferential price (FIT price) for rooftop solar power of 1,943 VND/kWh, equivalent to 8.38 cents/kWh, which would last for 20 years, while the initial FIT rate was 9.35 cents/kWh. Although there was a slight decrease in the preferential prices between the two phases, with relatively high profit and short payback period, many households still decided to invest in rooftop solar power. By the end of December 31, 2020, 101,029 rooftop solar power projects had been connected to the power system with a total installed capacity of up to 9,296 MWp. This was a good sign for the future of green energy in Vietnam. By early 2021, the total installed capacity of solar power nationwide had reached about 19,400 MWp (of which nearly 9,300 MWp is rooftop solar power), equivalent to about 16,500 MW, accounting for about 25% of the total installed capacity of the national power system. According to statistics of EVN, the entire electricity output generated from solar power nationwide in the whole year of 2020 was 10.6 billion kWh (of which, rooftop solar power alone is 1.15 billion kWh), accounting for about 1.15 billion kWh, equivalent to 4.3% of the total mobilized output of the entire national power system.

Vietnam is a country with great potential for solar energy, but it is unevenly distributed, with the majority concentrated in the southern region. Therefore, in addition to solar power projects in the South, power supply projects in the North will also be given priority in the coming period, although the electricity output in the North is not as high as the South. The sun can well meet the daytime peak load, contributing to reducing the demand for electricity transmission from the South to the North. However, by 2025, due to low reliability, the flexible power sources, including ICE engines, single-cycle gas turbines using LNG, and energy storage systems, will be essential for the electrical power system to ensure backup for solar power specifically and renewable energy in general. In addition, the government needs to develop a mechanism for pricing backup capacity for flexible power sources, to ensure their financial viability due to the low number of generating hours.

Currently, solar power is being intensively harnessed both on land, on water surfaces, and even on rooftops. By 2045, the total output of large-scale ground-mounted and grid-connected solar power is estimated at about 2.3 PWh/year; solar power on water surfaces would reach about 123 TWh/year; while rooftop solar power will have an output of 76 TWh/year (PDP VIII). However, the rapid growth of solar power in Vietnam also brings about many consequences. The Government needs to have appropriate policies to ensure stable and sustainable energy security.

## *4.2 Wind Power*

Besides solar power, wind power is also a clean and renewable energy source, which has been gaining much attention both domestically and internationally. Major changes have been made compared to the revised Power Master Plan VII (approved in 2016). In the period up to 2030, the power source development program of Power Master Plan VIII states wind power will develop on a large scale with power capacity increased by three times. Specifically, the installed capacity in 2020 was 630 MW, while the forecasted number in 2030 is 5,990 MW in PDP VII, and expected to increase to 18,010 MW in PDP VIII. In which, onshore and nearshore wind power will add 9GW, while offshore wind is expected to contribute an additional 2–3 GW.

Vietnam has considerable potential for wind power in the South, South Central, and Central Highlands regions, while the potential in the North is not great. In 2023, the Northern region would begin receiving electricity distributed from the North Central and Central power system.

The way wind turbines generate electricity is based on a very simple principle: The energy of the wind causes 2 or 3 blades to rotate around a rotor. The rotor is connected to the main shaft and the main shaft will drive the generator to rotate the generator to convert wind energy into mechanical energy and generate electricity. Wind turbines are usually placed on tall towers, about 30m or more so that they can capture wind at high speeds and in stable conditions. Output power produced by wind turbines will be determined based on the size of turbines. Thus, the bigger the rotor, the greater the ability to produce electricity and promote the effect of increasing wind power turbines. When wind power produces electricity, the grid can blend into the transmission countries, and distribute to the place of consumption. Wind power is a type of energy that has a lot of potential in the future such as:

- Theoretical potential
- According to Circular 06/2013/TT-BCT, "Theoretical wind power potential" is the wind power potential determined with wind speed of 6.0 m/s or more at an altitude of 80 m. The international theoretical potential assessment, meanwhile, uses a wind speed of 4.5 m/s. However, most of the suitable locations are located along the South Central Coast, the South, the Central Highlands, and the North Central region.
- Technical potential

- Areas having technical potential are those with the theoretical potential but considered able to build and operate wind power plants with existing technical conditions and technology.
- Regarding onshore wind power, technical potential is determined based on theoretical potential with current status and land use planning. On the basis of overlaying maps to eliminate areas with theoretical potential but not suitable for wind power development, the total technical potential of the whole country can reach 42,608 km$^2$, equivalent to 217.3 GW. Excluding the low wind speed of about 5.5 m/s, which is considered unfeasible with the current level of science and technology, the technical potential for wind power on land is about 47 GW. The results of calculation of technical potential of wind power on land by regions are as follows (Table 4):
- Regarding offshore wind power, according to the results of another study by the Danish Energy Agency (DEA) for the Ministry of Industry and Trade in 2020, the technical potential is about 162 GW. In which, the offshore wind area with fixed foundation (seabed depth ≤ 50 m) is about 132 GW and floating foundation is about 30 GW (Table 5).
- Through analysis and calculation, the technical potential for offshore wind power can reach 31,808 km$^2$ or 162,200 MW (Nguyen, 2022).
- Planning potential (or economic potential)
- A land area can serve different functions. Normally, if planned for one specific purpose, it would no longer be suitable for others. However, in the case of wind power, it should be noted that wind turbines in an area will have to be apart from others to avoid turbulence. The permanently used area would be about 5% of the study area, and the remaining of approximately 95% is still entirely suitable for other purposes, such as livestock farming, or even forestry, and can be integrated with wind power to promote effective land use and exploitation of efficient natural resources.
- Financial potential (or potential feasibility according to the current wind power tariff)

**Table 4** Technical potential of wind power on land by regions

| | Potential (km$^2$) | | | | Potential (MW) | | | |
|---|---|---|---|---|---|---|---|---|
| Wind speed range (m/s) | Northern | Central | Southern | Total | Northern | Central | Southern | Total |
| 4.5–5 | 1,898 | 11,807 | 2,955 | 16,660 | 9,679 | 60,218 | 15,069 | 84,966 |
| 5–5.5 | 473 | 8,139 | 8,141 | 16,752 | 2,412 | 41,598 | 41,517 | 85,437 |
| 5.5–6 | 83 | 4,262 | 3,287 | 7,632 | 421 | 21,738 | 16,764 | 38,923 |
| 6–6.5 | 7 | 1,187 | 56 | 1,250 | 37 | 6,051 | 284 | 6,373 |
| 6.5–7 | 3 | 238 | 0 | 241 | 15 | 1,213 | 0 | 1,228 |
| 7–7.5 | 0 | 59 | 0 | 59 | 0 | 303 | 0 | 303 |
| >8 | 0 | 14 | 0 | 14 | 0 | 72 | 0 | 72 |
| Total | 2,464 | 25,707 | 14,438 | 42,608 | 12,564 | 131,103 | 73,635 | 217,305 |

Source: The PDP8 project team

**Table 5** Technical potential of offshore wind power by regions

| | | Area (km²) | | | | | Technical potential (MW) | | | | |
|---|---|---|---|---|---|---|---|---|---|---|---|
| No. | Region | 6–6.5 m/s | 6.6–7 m/s | 7.1–7.9 m/s | >8 m/s | Total | 6–6.5 m/s | 6.6–7 m/s | 7.1–7.9 m/s | >8 m/s | Total |
| I | Northern | 1,578 | 971 | – | – | 2,549 | 8,048 | 4,952 | – | – | 13,000 |
| II | North Central | 607 | 266 | 107 | – | 980 | 3,095 | 1,357 | 545 | – | 5,000 |
| III | South Central | 4,671 | 2,000 | 4,540 | 11,930 | 23,141 | 23,821 | 10,200 | 23,154 | 60,841 | 118,000 |
| III | Southern | 3,808 | 1,330 | – | – | 5138 | 19,419 | 6,783 | – | – | 26,200 |
| Total | | 10,664 | 4,567 | 4,647 | 11,930 | 31,808 | 54,383 | 23,292 | 23,699 | 60,841 | 162,200 |

Source: C2Wind (2020)

- This is an area that can be effectively exploited when the project is run, bringing profits to the investor in the current conditions. As a result, the financial potential can change greatly over time. An area that is considered to have no financial potential at the moment, but may have this potential next year.

## 4.3 Biomass Power

Finding and exploiting renewable energy sources is a new direction to solve the energy problem and reduce environmental pollution caused by mining and using fossil fuels. Biomass fuel is considered a renewable but unsustainable form of fuel, because it is created mainly from plant materials and agricultural, forestry, and fishery wastes. The advantages of biofuels compared to fossil fuel combustion is not producing $SO_2$, a very toxic gas, causing environmental pollution. More importantly, generating energy from biomass does not pose a safety risk as with nuclear fuel or oil. Therefore, many countries, including Vietnam, have policies to encourage the use of biomass fuel.

Vietnam is an agricultural country with three consecutive rice crops in a year. Every year, the amount of waste generated during processing agricultural products and food is very large and diverse in type and quality. Therefore, the available supply of biomass fuels is abundant, diversified, and cost-effective. As a result, biomass is being considered as a preferred alternative fuel source both now and in the future.

Evaluation on biomass sources includes current assessment and future potential consideration. Assessment of the current situation will be based on data from planning, development plans outlined in the Decision approved by central government, government departments, and local authorities related directly or indirectly to biomass supply–demand. The methodology evaluates the theoretical potential of future biomass resources by examining future planning, crop productivity, social issues, and environmental protection.

According to the draft report "National Biomass Power Development" by the Institute of Energy (2018, in Information and Communications Magazine, 2022), by 2025, the theoretical potential of biomass sources in Vietnam would increase by 1.9%/year, reaching 130.69 million tons (equivalent to 454.89 million MWh). For the period from 2026 to 2030, the expected growth rate is 0.6%/year, reaching 138.41 million tons by 2030 (equivalent to 483.16 million MWh). Agricultural byproducts still account for a large proportion with over 67%, followed by wood with about 30%, and the rest is wood waste with about 3%.

Currently, the domestic biomass power generation potential is quite high with the main materials including wood, bagasse, and other biomass, which will reach a total capacity of 5316 GW, equivalent to a total power output of nearly 31 TWh. Bagasse and firewood are likely to be the sources of the biomass exploited with largest electricity generation, followed by rice husks and straw resources. In addition, Vietnam's biomass energy is being exploited, produced, and exported to other countries such as Japan, Korea, or European countries. This creates a significant additional source of income for farmers in our country.

## 4.4 Geothermal Power

Along with energy sources from wind, sun, ocean waves,. geothermal is a green, clean and environmentally friendly energy source used by many countries around the world. Geothermal energy is the energy generated from the thermal heat in the Earth. This energy is derived from the early formation of the planet, from radioactive decay of minerals, and from solar energy absorbed at the Earth's surface. The geothermal energy source in the ground is endless, ensuring that the geothermal power plant can operate sustainably for a long time. Moreover, a geothermal power plant also takes up very little space. Geothermal power plants do not burn any fuel, so they are cleaner for the environment than any other power plant. Geothermal resources by themselves cannot fundamentally solve energy problems, but make a small contribution to reducing dependence on the use of fossil fuels.

In Vietnam, the study of geothermal energy started from the 1980s and 1990s, however, the current applications of this energy are primarily for direct use, such as bottled mineral water, hot baths for medical treatment, hot spring resorts (e.g., in Binh Chau), drying of agricultural products, and production of iodized salt and $CO_2$ gas. Meanwhile, as a power source, its potential still remains untapped. The advantages in geothermal resources are evenly distributed across the territory, allowing widespread use in many localities. However, in areas with low temperatures, generating electricity can be costly.

Currently, there are five different types of geothermal energy sources, two of which, hydrothermal reservoirs and earth energy, have been put into commercial exploitation. The use of the others including geopressurized brine, hot dry rock, and magma still requires the development of advanced technology. Geothermal power plants work based on three main techniques in the production of electricity from geothermal that are technical dry steam, technical steam, and engineering binary cycle. According to the assessment of the experts and researchers in Vietnam for the development of geothermal power, there are several areas with potential for development of small-capacity power plants. Here is a summary of the potential by regions (Table 6):

The potential scale of development of geothermal power is very large, the theoretical potential of geothermal energy source locations has a total capacity of 696 MW with the power output of 3.45 TWh/year. Meanwhile, in terms of technical potential, the total installed capacity is 461 MW with the generated electricity output of about 2.28 TWh/year. The current geothermal plants are not large scale.

**Table 6** Synthesis of geothermal potential by regions

| Geothermal zone | Number of sources | Temperature (°C) | Number of potential geothermal sources by scale | | |
|---|---|---|---|---|---|
| | | | Industrial | Medium | Small |
| Northwest | 79 | 103–200 | 10 | 25 | 44 |
| Northeast | 11 | 95–146 | 2 | 6 | 3 |
| Red River Delta | 17 | 100–150 | 5 | 3 | 9 |
| North Central | 42 | 120–210 | 4 | 10 | 28 |
| South Central | 67 | 110–200 | 14 | 18 | 35 |
| Southern region and Mekong River Delta | 53 | 150 | | 22 | 31 |
| Total | 269 | | 35 | 84 | 150 |

Source: Vietnam Institute of Geology and Minerals

## *4.5 Small Hydrogen Power*

The trends in developing wind energy and solar energy are growing. However, that the solar plants cannot generate electricity when there is no sunlight, and the wind turbines cannot operate when there is no wind, causes a serious shortage of capacity on the national grid. Moreover, the rate of load change of these renewable energy sources is also quite fast, making the operational role of hydropower, including small hydropower, even more important. With the high growth of electricity consumption and the forecast of energy shortage in the future, the potential of small hydroelectric power is an energy source that needs to be explored, exploited, and developed sustainably for economic and socioeconomic development, as well as energy security. Meanwhile, some large hydropower projects are expected to be depleted in the future, making the problem of energy supply increasingly urgent.

Unlike large hydropower, small hydropower has a smaller scale, creating less environmental and social impacts. Thus, it is classified as a renewable energy source. Small hydro is the development of hydroelectric projects at a scale appropriate for local communities and industries, or contributing to distributed electricity generation in the regional grid, with a capacity of less than 50 MW can be further subdivided. In small hydroelectric projects, the scale of the work typically involves a low dam, a small path, limited construction volume, and minimal land use. Each small hydropower station usually has only 2–3 units, transformers, power distribution stations and 35 kV or 110 kV power transmission lines. Small hydropower plants, if they have a reservoir, usually have a small capacity, or they can operate without one.

The total small hydropower potential (under 30 MW) of the whole country is about 10,000 MW. However, due to the impacts of small hydropower on the environment and forest conservation, the Department of Electricity and Renewable Energy has reviewed the planning and potential locations, and about 4,000 MW were eliminated. Therefore, only about 6,000 MW of small hydroelectricity can be exploited, of which, output in the North is about 3,000 MW, in the Central Highlands region is 1,000 MW, in the North Central, Central Central, and South Central regions have about 600 MW/region, and about 75 MW in the South. Currently, the total capacity of small hydropower sources that have been built and put into operation is more than 3,200 MW (PDP VIII). So basically, hydropower has been exploited most of its potential.

Most of the remaining exploitable capacity has been researched and invested in. Those projects are either in the preconstruction stage or already built. According to the progress of hydroelectric project implementation, the system can add about 2.8 GW of small hydro capacity in the period 2020–2025, and approximately 2,700 MW by 2030.

Compared to hydroelectric power plants, the technological trend of small hydropower plants has hardly changed. Technological and general equipment using old technologies still accounts for 21% of the turbine hydraulic and 23% of the generator power. Small hydro turbines currently limit hydraulic efficiency to between 90% and 95% with the average Tmax at about 3000 h. Spillway technology is increasingly used for small hydroelectric plants, with materials such as, homogeneous earth dams, rockfill dams with clay cores. By 2030, an additional 44,983 ha of land will be mobilized for hydropower and small-scale hydropower projects. Thus, it is forecasted that the total accumulated land area by 2030 for the entire power sector would reach 124,501 ha. The cost of small hydropower is highly dependent on location, topography, and other natural conditions, so it is difficult to predict a standard output value for estimating technical and economic parameters.

- Investment rate: Investment rate for small hydroelectricity in Vietnam fluctuates greatly. According to statistics based on the database of 120 projects, the average investment rate is 850 USD/kW. Since small hydropower plants (under 30 MW) that will have been put into operation before 2030 all located in difficult areas for construction and operation, the expected investment rate for small hydropower is about 1,200–1,500 USD/kW (not yet available), in which, equipment accounts for 25%–30%, while construction and installation accounts for 70%–75%.
- The average capacity factor of 120 small hydropower projects currently in operation is 56%, and is expected to remain unchanged for future projects by 2030.
- Charged cost: for small hydropower plants, the "avoidable cost" mechanism will be applied. (PDP VIII)

Resolution 55-NQ/TW dated February 11, 2020 emphasizes that small hydropower development must be selective, together with building a long-term electricity import strategy. Due to the above-mentioned features, it is necessary to ensure technological-economic and land-use efficiency, as well as environment protection, especially natural and protection forest reserve while planning and constructing (Nguyen, 2024). By 2045, to provide additional backup for the dry scenario, the government may consider adding 0.9 GW of small hydropower capacity.

## References

### *Legal Documents*

C2WIND (2020) Vietnam offshore wind - country screening and site selection

Circular No. 06/2013/TT-BCT dated March 08, 2013 of Ministry of Industry and Trade on regulation on content, order, procedures for formulation, appraising and approving wind power development planning

Conclusion 26-KL/TW dated October 24, 2003 of the Politic Bureau on Strategy and planning for developing Vietnam's electricity

Decision 839/QD-TTg dated July 26, 2023 by the Prime Minister on Approving the National Energy Master Plan for the 2021 – 2030 period, with a vision to 2050

Decision No. 13/2020/QD-TTg dated April 06, 2020 of the Prime Minister on mechanisms to promote development of solar power in Vietnam

Decision No. 39/2018/QD-TTg dated September 10, 2018 of the Prime Minister on amending several articles of Decision no.37/2011/QD-TTg dated June 29, 2011 of the Prime Minister on provision of assistance in development of wind power projects in Vietnam

Notice No. 3116/TB-TTCP dated December 25, 2023 of the Government Inspectorate concluding the compliance with policies, laws regarding PDP VII and revised PDP VII

Resolution 18-NQ/TW dated October 25, 2007 of the 10th Political Bureau on strategic orientation for Vietnam's national energy development to 2020, vision to 2050

Resolution No. 55-NQ/TW dated February 11, 2020 of the Politburo on Orientations of the Vietnam's National Energy Development Strategy to 2030 and outlook to 2045 (Resolution 55)

Updated Nationally Determined Contribution (NDC) (2020). Available at: https://unfccc.int/sites/default/files/NDC/2022-06/Viet%20Nam_NDC_2020_Eng.pdf

Vietnam's National Power Development Plan of 2021–2030 and Vision to 2050 (PDP VIII)

Vietnam's Revised National Power Development Plan 7 for the period 2011–2020 with a vision to 2030 (PDP VII Rev)

### *Others*

Danish Energy Agency (2017) Vietnam energy outlook report 2017. https://www.ea-energianalyse.dk/wp-content/uploads/2020/12/1639_vietnam_energy_outlook_2017.pdf

Duy D (December 22, 2020) Giải bài toán đảm bảo an ninh năng lượng và phát triển bền vững (Solutions to the problem of ensuring energy security and sustainable development). Vietnam Plus. Retrieved from: https://www.vietnamplus.vn/giai-bai-toan-dam-bao-an-ninh-nang-luong-va-phat-trien-ben-vung/683597.vnp

EVN Department of Communication (2024). *Một số số liệu tổng quan về nguồn điện toàn quốc năm 2023 (Some data on national energy supply in 2023)*. EVN. Retrived from: https://www.evn.com.vn/d6/news/Mot-so-so-lieu-tong-quan-ve-nguon-dien-toan-quoc-nam-2023-66-142-124707.aspx

Huong T (January 24, 2015) Khai thác dầu khí có những biến động mới (Update in oil and gas exploitation). Đầu tư online (Investment Online). Retrieved from: https://baodautu.vn/khai-thac-dau-khi-co-nhung-bien-dong-moi-d242373.html

Huong, T. (Jan 23, 2025). Năm 2024, *Việt Nam nhập khẩu 63.8 triệu tấn than (In 2024, Vietnam imported 63.8 milllion tons of coal).* Đầu tư online (Investment Online). Retrieved from: https://baodautu.vn/nam-2024-viet-nam-nhap-khau-638-trieu-tan-than-d242396.html

Nguyen AT (August 15, 2022) Điện gió ngoài khơi: Tiềm năng, Thách thức, và Kế hoạch hành động của Việt Nam (Offshore Wind Power: Potentials, Challenges, and Action Plan of Vietnam. Năng lượng Việt Nam (Vietnam Energy). Retrieved from: https://nangluongvietnam.vn/dien-gio-ngoai-khoi-tiem-nang-thach-thuc-va-ke-hoach-hanh-dong-cua-viet-nam-29248.html

Nguyen H. H. (2024). Thuỷ điện Việt Nam - Tiềm năng còn lại (có khả năng khai thác) và câu hỏi còn để ngỏ (Hydropower in Vietnam – Available potential (exploitable) and the unanswered question). Năng lượng Việt Nam (Vietnam Energy). Retrieved from: https://nangluongvietnam.vn/thuy-dien-viet-nam-tiem-nang-con-lai-co-kha-nang-khai-thac-va-cau-hoi-con-de-ngo-33369.html

Phan MH (February 05, 2024) Vietnam should tackle pressure from carbon tax head-on. RMIT University. Retrieved from: https://www.rmit.edu.vn/news/all-news/2024/feb/vietnam-should-tackle-pressure-from-carbon-tax-head-on

Tiềm năng điện sinh khối cho sự phát triển bền vững ở Việt Nam (Potential of Biomass Energy for Sustainable Development in Vietnam) (November 17, 2022) Tạp chí Thông tin và Truyền thông (Information and Communications Magazine). https://ictvietnam.vn/tiem-nang-dien-sinh-khoi-cho-su-phat-trien-ben-vung-o-viet-nam-54176.html

**Hanh-Thi Vu** (PhD in Economics and Management, Université Libre de Bruxelles, Belgium) is a lecturer in International Business, Foreign Trade University, Vietnam. Her research interests include sustainable trade, international business, and waste management.

**Ha-Ngoc Van Nguyen** received her Bachelor's degree in Economics from the Foreign Trade University, Hanoi, Vietnam. She is working for GIZ Energy Support Program, Vietnam. Her research interests relate to sustainable development, waste management, and international trade.

# Identification of Wetland Goods and Services in Vietnam: The Case of Cam Thanh Wetland

**Trang-Huyen Le and Thuy Anh-Tu**

**Abstract** Human impacts on wetlands have been ongoing for decades, changing both biodiversity and the cover of wetlands. The wetland goods and services assessments aim to capture significant social-ecological benefits and integrate those benefits into wetland management and human well-being. The existing approaches to identifying goods and services of wetlands are primarily based on the wetland functions and values. In this chapter, we introduce a new approach to identifying goods and services of wetlands for Vietnam: a stakeholder approach. Without loss of generality, this new methodology is presented in the Cam Thanh commune, Hoi An City, Quang Nam province case. Researchers can easily apply the same method to different locations in Vietnam to draw relevant conclusions. From the study site employed to examine goods and services of the wetland ecosystem in Cam Thanh to illustrate a new approach, it is interesting that a new wetland service was contingently identified. It is seen that the utilization of wetland goods and services play a fundamental role in providing more information and enabling pressures to be integrated into environmental policy and wetland management.

## 1 Introduction

Wetlands are one of the ecosystems that provide numerous biological and economic benefits to humans (Murat and Ahmed 2023). However, wetlands around the world have been degraded, lost or are under serious threat all over the world.

T.-H. Le (✉)
Faculty of International Economics, Foreign Trade University, Dong Da, Ha Noi, Vietnam
e-mail: tranglh@ftu.edu.vn

T. Anh-Tu
Journal of International Economics and Management, Foreign Trade University, Dong Da, Ha Noi, Vietnam
e-mail: thuyanh.tu@ftu.edu.vn

 
H.-T. Vu, L.-P. Dana (eds.), *Vietnam on the Road to Sustainable Economic, Social and Business Development*, Frontiers in South and Southeast Asian Development Research, https://doi.org/10.1007/978-981-96-4371-4_3

Although there is an increasing recognition of the need to conserve wetlands and many international agreements and national policies have been enacted, the loss of wetlands continues to occur. For sustainable management and optimal management policies of wetlands, it is useful to provide information on the economic value of wetlands from stakeholders. Information on the economic value of wetland goods and services provided through stakeholder assessments offers comprehensive and in-depth insights to decision-makers in wetland management and use.

Due to the highly dynamic nature and difficulty in defining wetland boundaries, there is much debate among scientists about the definition of wetlands. Governments and nongovernmental organizations widely accept the definition of a wetland in the Ramsar Convention as "*areas of marsh, fen, peatland or water, whether natural or artificial, permanent or temporary, with water that is static or flowing, fresh, brackish or salt, including areas of marine water, the depth of which at low tide does not exceed 6m*" (Springgay and Janzen, 2019). This definition conveys many important characteristics of wetlands and implies their complexity. However, this definition does not provide any guidance on the general characteristics of wetlands to indicate the actual functions of wetlands. The characteristics of wetlands show that wetlands are a natural resource that provides important services to support the lives of humans, animals and plants. Wetlands can be found in marsh, wet meadows, wet prairie, swamp, bog, fens and seeps, vernal pools, and prairie potholes. Another definition of wetland "are lands transitional between terrestrial and aquatic systems where the water table is usually at or near the surface or the land is covered by shallow water. Wetlands must have one or more of the following three attributes: (1) at least periodically, the land supports predominately hydrophytes; (2) the substrate is predominately undrained hydric soil; and (3) the substrate is nonsoil and is saturated with water or covered by shallow water at some time during the growing season of each year" (Cowardin et al., 1979).

The definitions used in this chapter are as follows: "A wetland is an ecosystem that depends on constant or recurrent, shallow inundation or saturation at or near the surface of the substrate. The minimum essential characteristics of a wetland are recurrent, sustained inundation or saturation at or near the surface and the presence of physical, chemical, and biological features reflective of recurrent, sustained inundation or saturation. Common diagnostic features of wetlands are hydric soils and hydrophytic vegetation. These features will be present except where specific physicochemical, biotic, or anthropogenic factors have removed them. or prevented their development."

At the international level, wetland conservation is promoted through an intergovernmental agreement, the Ramsar Convention, which came into force in 1975. The Ramsar Convention provides a framework for international cooperation to ensure the conservation of wetland habitats, sustainable management, and use of wetland resources. At the national level, wetland conservation is carried out by establishing national parks and nature reserves. In addition, the publication of national Redbooks—lists of threatened species—is another measure adopted by governments and non-governmental organizations to help protect species that inhabit wetlands from the risks posed by environmental changes.

In many countries worldwide, wetland management has been integrated into regional, interregional, and national socio-economic development plans. Previous economic development strategies that prioritized converting wetlands into industrial zones have gradually shifted toward policies that combine economic growth with wetland conservation. These newer approaches recognize the importance of avoiding the fragmentation of wetland ecosystems for development purposes, in order to maintain or restore the full range of wetland functions—such as water filtration, aquatic habitat restoration, and climate regulation.

However, the international agreements and regulatory frameworks are not sufficient. Local residents have used their right to improve their living conditions and have yet to consider the effects on a broader geographical scale, which can be seen as interrelated market and intervention failures. The reason is that policymakers, conservationists, and the general public do not frequently underestimate the role and function of wetlands and the indirect effects of wetlands on land use and environmental management.

The underestimation of the economic function and role of wetlands from human activities stems from the complexity and often invisible interconnections within wetland ecosystems. These include spatial relationships between vegetation, surface water, and groundwater. As a result, many human activities—such as industrial wastewater discharge or the use of agricultural chemicals—may appear to have only localized, surface-level impacts. However, they can lead to broader environmental consequences, such as groundwater contamination or pollution of distant water sources. These impacts often cause economic losses to affected communities, who bear the consequences but receive no compensation from the polluters.

Wetlands are often viewed as natural waste receptacles for various sources of pollution, such as waste from residential areas, industrial plants, and agriculture, which are important factors causing wetland pollution and degradation. Many wetlands with essential functions, such as water provision for daily life, production process, and irrigation, are often viewed as open-access resources. Hence, the lack of property rights leads to the inability to limit the unrestricted exploitation of resources to the exhaustion.

The above analysis showed the need to determine the value of wetland ecosystem functions. Research on ecosystem goods and services was carried out in the late 1960s, defining and classifying the concept. Many studies aim to estimate the value of ecosystem goods and services according to the ecosystem approach to give more weight to the ecosystem system in policy decisions. The benefits of ecosystem functions transferred into the value of ecosystem goods and services are carried out to increase human interest in conserving biodiversity (Helliwell, 1969). Although ecosystem services in developing countries have declined sharply, many residents still depend directly on wetland ecosystem services for their livelihoods. The integration of ecosystem goods and services into national environmental policies was formally adopted at the Earth Summit in Rio in 1992, as a foundation for the concept of biodiversity in the ecosystem approach (Lafoley et al., 2004).

The identification and definition of ecosystem goods and services are considered a complex field, requiring scientists from several disciplines to work together:

ecologists, economists, modelers, social scientists, policymakers, and ecosystem managers (Finisdore et al., 2020).

However, more research is needed to identify the ecosystem goods and services based on the expression of utilities of humans who benefit from ecological systems. This chapter aims to define goods and services of wetlands based on the interpretation and assessment of direct, and indirect utilities as perceived by beneficiaries or related stakeholders, providing a foundation for estimating the economic value of wetlands.

Two independent activities were conducted to achieve a precise and holistic identification of wetland goods and services, enabling a comparison between theoretical concepts and real-world observations. The first involved reviewing theories from published papers and books on wetland ecosystem goods and services. The second consisted of on-site examinations of the goods and services provided by wetlands. To present a holistic identification of what wetlands offer to humans, this chapter illustrates the benefits of a specific wetland by considering the utilities perceived by both users and non-users.

## 2 Framework to Identify and Define the Concept of Wetland Goods and Services in Vietnam

### 2.1 Stakeholder Approach

Within the context of the broader definition of stakeholders, stakeholder engagement is defined as "*the process used by an organization to engage relevant stakeholders for a purpose to achieve accepted outcomes.*"

Community-based wetland management is a people-centered approach to management. Wetlands play an important role in the prosperity and well-being of people who rely on wetland resources as a sole source of income and food. The participation of stakeholders in the management of natural resources is a prerequisite for sustainable resource use. Concerning wetlands, the Ramsar Convention recognizes participatory management as a fundamental principle of wise use.

There are many different definitions of stakeholders, according to the World Health Organization, stakeholders "*are persons, groups or institutions with interests in a project or policy or who may be directly or indirectly affected by the process or the outcome*" (WHO, 2005). The stakeholders in stakeholder-based wetland management are defined as individuals or organizations that support or oppose a project that is likely to affect wetlands, by which definition, stakeholders (i) are likely to be affected by wetland management, (ii) will be involved in implementing wetland management activities, or (iii) support or oppose a research project or development activity or policy on wetlands. Thus, NGOs, governments, scientists, policymakers, researchers, private businesses, financial institutions, and people who use wetlands are all stakeholders involved in wetland management.

Stakeholder involvement in the wetland resource management process is generally expected to result in the following outcomes: (i) a better understanding of people's concerns, and thus solutions that are more relevant to their needs; (ii) an appreciation of people's knowledge of wetland systems, the integration of this knowledge into wetland management options, and better targeting of awareness raising and education activities; (iii) ownership of the project and support for project implementation; (iv) reduction of potential conflicts among stakeholders; and finally (iii) improved communication and coordination of actions and enhanced working relationships among stakeholders.

Grimble and Wellard define stakeholder analysis as "a *holistic approach or procedure for understanding of a system, and assessing the impact of changes to that system, identifying the key actors or stakeholders and assessing their respective interests in the system*" (Grimble and Wellard, 1997). The World Health Organization defines stakeholder analysis as "a technique used to identify and assess the importance of stakeholders, who may be affected by an activity, a project, a policy" (WHO, 2005). The World Bank defines stakeholder analysis as identifying those involved in, or affected by, or influenced by a community's activities and grouping them according to their level of participation or their interest or influence in the community. Thus, stakeholder analysis is a project management tool to determine the outcomes of implementing a particular project or policy, and stakeholder analysis is a tool to understand the complexity, development potential, and trade-offs between different objectives of a project or policy.

Identifying stakeholder groups is an essential and sensitive step that needs to be considered in a case-by-case situation, using specific criteria based on the purpose of the analysis. For example, based on the purpose of natural resource management, stakeholder grouping is based on the level of intervention criteria from global, national, and regional levels down to household or even intra-household levels. The grouping can refer to the classification of stakeholders to avoid missing important factors (Fig. 1 and Table 1).

In this chapter, we introduce a new stakeholder-based approach, to wetland goods and services at a selected site. This approach allows a combination of both ecological functions and total economic values. The research attempted to enumerate the maximum number of wetland stakeholders, which ensures the comprehensive assessment of wetland functions identified by ecological scientists and the values benefited by users and nonusers to avoid omitting any goods and services of the ecosystem (Fig. 2).

Without loss of generality, this new methodology will be presented in the case of Nipa palm wetland in Cam Thanh commune, Quang Nam province. Researchers can easily carry on other research using the same methodology applied to other locations to draw relevant conclusions.

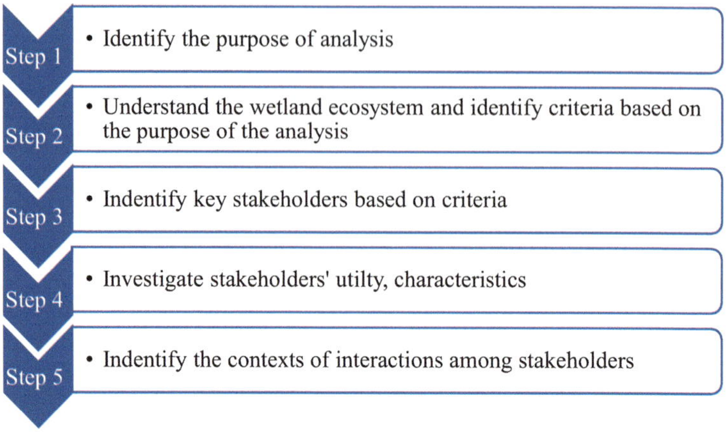


**Fig. 1** Steps to group stakeholders. Source: Authors

**Table 1** Grouping stakeholders by based on the level of natural resource management

| Top-down level | Stakeholders' example | Environmental utility |
|---|---|---|
| International | International environmental organizations<br>Multinational corporations<br>International and regional NGOs<br>Global nonprofits conservation, environmental protection organization | Biodiversity conservation<br>Climate change<br>Global resource reserves<br>Threats reduction to the diversity of life on Earth |
| Country | National NGOs<br>Policymakers<br>Donors | Logging<br>Landscaping for tourism<br>Climate regulation<br>Biodiversity conservation |
| Regional | Regional authority<br>Regional planner<br>Resource-based communities | Water supply<br>Erosion control<br>Resource exploitation |
| Local | Local authorities, local official<br>Residents: farmers, fuel collectors, exploitation company | Timber products<br>Cultivation, irrigation<br>Cultural and historical sites<br>Scientific research |

Source: Grimble and Wellard (1997)

## 2.2 *Types of Values*

Wetlands provide many essential goods and services to humans, which means that wetlands have many potential economic values but, at the same time, are ecologically sensitive and adaptive systems (Rai and Raleng, 2011). Why are these values often overlooked in the design and promulgation of wetland-related policies? In many cases, wetlands continue to be replaced by economic activities, and their degradation persists despite growing awareness.

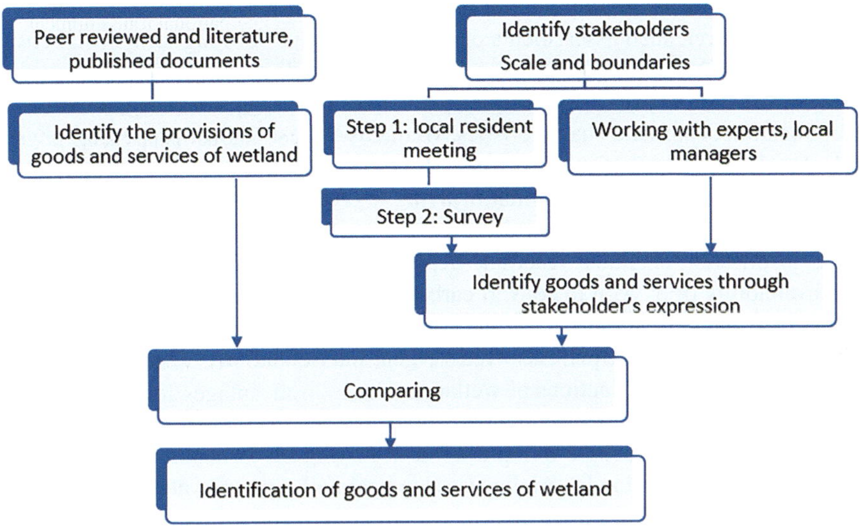


**Fig. 2** Framework to integrating stakeholders and wetland goods and services. Source: Compilation by the authors

Many coastal wetlands, such as coastal mangrove forests, have disappeared or been fragmented due to the conversion into shrimp farms or resorts. Although many countries have implemented to protect wetlands—recognizing their crucial role as habitats for migratory birds—these ecosystems continue to be degraded. Urban expansion, technological pollution, and the prioritization of economic activities such as aquaculture, industrial zone development, infrastructure projects, and local tourism have all contributed to the ongoing fragmentation and loss of wetland areas.

For example, the Zatok wetland in Koper, Slovenia is an important habitat for migratory birds, yet it is severely polluted by waste from surrounding urban and industrial areas. Significant portions of this wetland have been encroached upon for urbanization, industrial development, road construction, and tourism. Similarly, the ecologically significant Doñana wetland in Spain initially suffered from overextraction of water for strawberry farming and was later transformed into a tourist village located in the heart of the wetland. In the Netherlands, despite the Wadden Sea being a major hub for migratory birds from Scandinavia, economic pressures led the Dutch government to permit large-scale gas extraction in the area. In Ireland, vast peatland areas have been exploited for fuel to meet the population's increasing energy demands. The Aral Sea wetlands are suffering from extreme water stress and desiccation, as upstream water flows have been diverted to irrigate cotton fields. The Everglades in the United States are polluted by nutrient runoff from nearby sugarcane plantations, as well as the widespread use of fertilizers, herbicides, and pesticides. In Southeast Asia, many coastal wetlands have been rapidly converted into fish and shrimp aquaculture ponds. The examples clearly illustrate the conflicting interests of stakeholder groups across various levels. In many cases, land-use changes were considered optimal at the time, based on the highest immediate net benefit. However, such assessments often fail to account for long-term social costs and the loss of broader ecosystem services that wetlands provide.

The valuation of wetlands using a stakeholder-based approach begins with the identification of wetland characteristics. Wetlands consist of three properties: physical, chemical, and biological. The biological properties of wetlands are mainly concerned with maintaining the habitat of living organisms, including decomposition, transformation, production and storage of nutrients, and respiration. Wetlands undergo different chemical reactions depending on pH, salinity, redox reactions, and availability of oxygen to convert carbon, nitrogen, and sulfur. Physical characteristics of wetlands include maximum soil depth, buoyancy rate of vegetation, wetland depth from shore and pH, dissolved oxygen, and wetland size. The characteristics of wetlands shape the functions of wetlands.

The structure of wetlands includes elements of water, soil, plants, wildlife, microorganisms, and people living around or relying on wetlands for their livelihoods. Water in wetlands originates from rainfall, groundwater, and surface water circulating within the wetland ecosystem. Wetlands soil is a mixture of soil and water, providing a foundation for hydrophytic (water-loving) plants and playing a crucial role in supporting the ecological function of wetlands. Vegetation typically includes species adapted to saturated environments, such as reeds, mangroves, and nipa palms. Wetlands also serve as vital habitats for various wildlife species—both resident and migratory—including birds, fish, and shrimp, which rely on these ecosystems for feeding, shelter, and reproduction. In addition, wetlands are home to many microorganisms and bacteria that play a role in decomposing and cycling nutrients.

In contrast, wetland processes refer to the transformation of matter and energy within wetlands. These transformations result from the interaction among hydrology, geomorphology, soil, and vegetation. The interplay between wetland structure and processes gives rise to essential ecosystem functions such as providing food for human livelihoods, filtering water, retaining nutrients, and controlling floods.

This analytical framework follows the connection among wetland characteristics, structures, and processes that shape wetland functions and in turn, the goods and services provided to society. It also helps identify the corresponding economic values of wetlands. Moreover, this approach aligns well with the methodology endorsed by the Convention on Biological Diversity.

The valuation of wetlands is based on a human-centered approach, recognizing that the goods and services provided by wetlands contribute directly to human well-being. However, valuing wetlands through the lens of human well-being requires an analysis of individual and collective preferences, as the economic value of wetlands depends on what people derive from them in terms of utility and satisfaction. Economic value is defined as the benefit a user or nonuser receives from a good or service or a resource measured by what they are willing to give up to obtain it. Therefore, economic value is interpreted either as the benefit gained from an increase in goods and services or the cost (in value or utility) of sacrificing them. (Fig. 3).

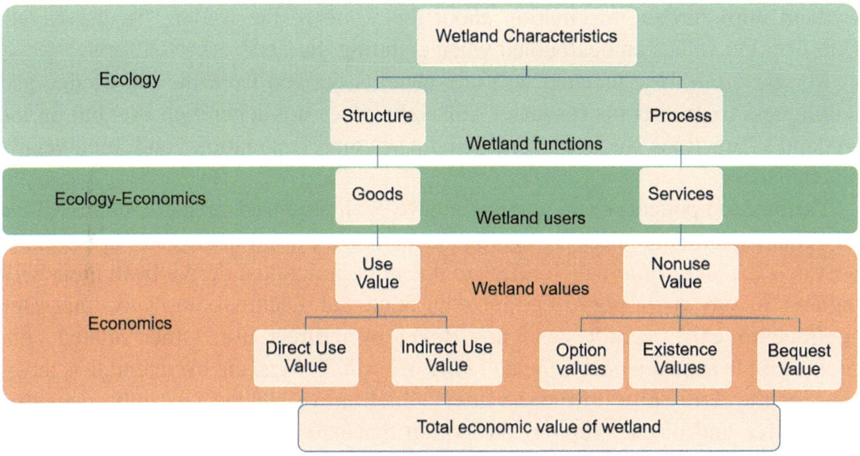


**Fig. 3** Ecological and economic values. Source: Turner et al. (2000)

The process of moving from the ecological characteristics of wetlands to their economic valuation of wetlands presents the fundamental link between wetlands function and their perceived value. The economic values of wetlands always depend on the functions that wetlands provide and are recognized by society as valuable. Therefore, the functions themselves may not have economic value if they are not recognized and exploited by people. These functions of wetlands only become economicaly significant when people recognize, and demand policies that support the goods and services provided by wetlands. For instance, wetlands serve as migratory habitats for birds which holds high natural ecological value. However this function may not yield economic based on anthropocentric criteria. Nevertheless, this function becomes economically valuable when conservationists are willing to invest in protecting these habitat. In this case, the function of the wetlands generates value by serving the interests of conservationists.

The total economic value (TEV) of a wetland is determined by the aggregate benefits received by individuals or by the total amount of resources they are willing to forgo in order to to improve or preserve the wetland. TEV can also be broken down into various components based on the direct or indirect benefits people receive from wetland goods and services.

To date, several terms have been found to represent the components of total economic value: *direct value* and *indirect use value*. Another type of wetland value often associated with wetland use value is *option value*, which is the value that individuals derive from preserving and protecting the resource so that it is available for future use. *Quasi-option value* is another approach to the economic value of wetlands, which is the value of the potential benefits derived from waiting for improved information before forgoing the option of preserving the resource for the future. This value depends on the change in preferences resulting from improved information prospects. For example, quasi-option value is the value of conserving a wetland until further information about the value of the wetland is discovered. Quasi-option value can be doubled when counting the TEV.

*Nonuse value* is associated with the benefits derived from the knowledge and willingness to protect the resource. This value does not depend on use, but on the wetland's functions such as bird habitat, climate regulation, and biodiversity maintenance.

Various components of nonuse value have been proposed in calculating the TEV of wetlands, although there is ongoing debate about these values. One such component is: *existence value* which refers to the value individual derive from their willingness to pay to preserve certain environmental features simply so that they continue to exist regardless of any direct or indirect use. Other related concepts include *instrumental value* and *intrinsic value*. Some environmentalists advocate for the term "pure intrinsic value," which move beyond the anthropocentric perspective and instead support the notion that nature possesses value in and of itself—independent of human use or benefit. This perspective introduces a distinction between the total social value of wetlands which encompass ethical, cultural, and ecological considerations—and the more narrowly defined total economic value, which focuses on human-centered valuations.

Economists have broken down the total economic value of wetlands into use values and nonuse values. So, who are the users and nonusers associated with wetlands and assign economic values to wetlands? Turner et al., (2000) have classified wetland stakeholders into nine groups (Table 2):

The benefits that stakeholders receive from wetlands can be either compatible or conflicting. Therefore, policymakers when implementing programs and projects related to wetlands need to consider the interests of stakeholders, evaluate the benefits they receive, and assess the trade-offs between groups to provide data for economic analysis, which then serves as useful information for the decision-making process. Economists have decomposed the total economic value conferred by resources into use values and nonuse value (Fig. 4).

**Direct use value (DUV)** is derived from using of goods and services of the environment for living, commercial and recreational purposes.

Direct use value refers to the goods and services of wetlands used for subsistence, commercial, and recreational purposes. Wetland goods used for consumption purposes such as vegetables for food, animals for food, aquatic products, seafood for food, wood for fuel, construction, and pharmaceuticals. Wetland services are used for recreational purposes such as fishing, picnicking, sightseeing, and jogging. Direct value is often exploited by groups of direct users, direct exploiters, and producers.

**Indirect use value (IDV)** relates to the situations when people get the benefits from environmental functions, usually measured by the ability to prevent environmental damages.

IDV are services provided by wetlands that are independent of the wetland's area, meaning that they benefit not only those living within and in the wetland's buffer zone, but also those living outside the wetland.

**Table 2** Wetland stakeholder groups

| Stakeholders | Description | Stakeholders' knowledge of wetland | Utility |
|---|---|---|---|
| *Direct extensive users* | Direct extractors of goods from wetlands, whose livelihoods depend on wetland restoration | Possess specific ecological knowledge within the context of their institutions and habitats | Harvest wood for fuel or timber<br>Logging for housing<br>Exploitation of plants and fruits for food<br>Hunting of birds and animals for food or pets<br>Exploitation of aquatic and marine products for food |
| *Direct intensive users* | Use new technology to exploit wetlands deeply | Have the ability to exploit wetlands to a large extent, even exceeding the productive capacity of wetlands, exceeding the resilience capacity and seriously reducing the quantity and quality of wetlands if not properly managed | Exploiting raw materials, fuels, and biomass of wetlands |
| *Direct exploiters* | Directly dredge sediments in wetlands or exploit minerals, peat, clay, sand | When exploiting, paying little or no attention to the reserves, resilience, and productivity of wetlands | Coal and peat mining<br>Mineral mining<br>Soil and sand mining |
| *Agricultural producers* | Convert wetlands into agricultural land | Exhausting resources within wetlands for agricultural purposes | Soil fertility and nutrient richness<br>Availability of water for irrigation purposes |
| *Water abstractors* | Exploit water resources for living purposes, agricultural irrigation, waste discharge, etc. | Reducing water levels, polluting or degrading water resources | Use water sources for drinking, washing, sanitation<br>Irrigation<br>Dumping or discharging solid waste, waste water into wetlands |

(continued)

**Table 2** (continued)

| Stakeholders | Description | Stakeholders' knowledge of wetland | Utility |
|---|---|---|---|
| *Human settlements close to wetlands* | Convert wetlands into residential areas and infrastructure | Continue to expand housing and public services by encroaching on wetlands | Use of utilities from wetlands Use of water for domestic use and production<br>Exploit resources for food and medicine<br>Dump wastewater and waste into wetlands |
| *Indirect users* | People who benefit from wetland services such as storm mitigation, flood mitigation, hydrological stabilization, and water purification | Few people may be aware of these benefits from wetlands because these services are provided to many people and are realized through wetland functions, so few people know about the source of these benefits | Reduced storm damage<br>Reduced flood damage<br>Higher productivity when using clean, nutrient-retaining, and stable water |
| *Nature conservation and amenity groups* | Taking action to conserve nature for the joy of the presence of animals and plants | Having an understanding of the value of animals and plants, the value of the landscape that these groups want to preserve | Satisfaction that the animals, plants, and natural landscape that they want to protect will still be present |
| *Nonusers* | People who recognize the intrinsic value of wetlands even if they have never visited the area | Understanding of the intrinsic value of wetlands is not related to whether or not they use any goods or services from wetlands | Satisfaction that the wetland values they want to protect will be retained |

Source: Turner et al.,(2000)

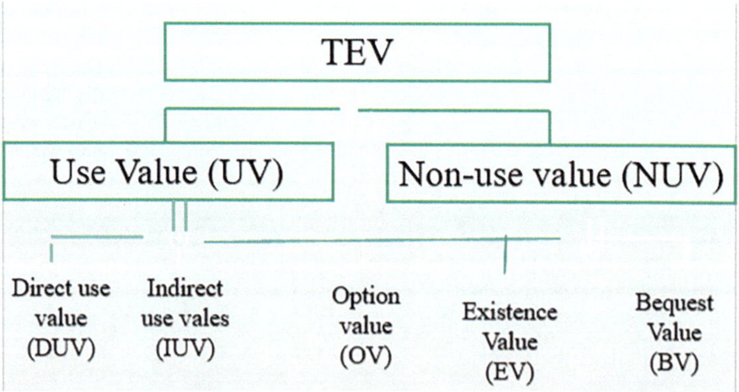


**Fig. 4** Total economic value. Source: Turner et al. (2000)

For example, wetlands' natural water filtration capacity benefits people downstream, the storm protection function of mangroves benefits coastal assets and infrastructure, and the carbon sequestration of wetlands contributes to reducing carbon emissions, thereby contributing to climate change mitigation.

**Option value** refers to situations where individuals are willing to pay to preserve environmental components or systems for future use.

Wetland goods and services such as provisioning, regulating, and culture are chosen for use now or in the future. Option value arises from the choice in the present to preserve wetland goods and services for future use.

**Existence values** are the specific values people attach to goods or services, regardless of how the goods or services are used. These values are often associated with habitats and biodiversity.

Existence values refer to the satisfaction people get from knowing that a resource they care about existence even if they do not expect to use it directly or indirectly. Existence values are sometimes referred to as passive use values.

**Bequest value** is the value that is retained for future generations.

Many individuals in today's society ensure the availability of biodiversity, goods and services of wetlands for future generations. The concerns and actions of the present generations have demonstrated to allocate resources to future generations. The perspective of these individuals is that the rights and opportunities of future generations to access resources are no less than those of the present generation. Therefore, understanding of the benefits of wetlands by the present generation will help preserve, protect and pass on these benefits to future generations better.

## 3 Goods and Services Identification of Nipa Palm Wetland in Cam Thanh Commune, Hoi An city, Quang Nam Province

### *3.1 Description of the Cam Thanh Wetland*

Nipa palm (Nipa fruticans) wetland known as Bay Mau nipa palm forest is located in Cam Thanh commune, Hoi An city, Quang Nam province, Viet Nam. Its name originates from the original forest area "Bay" meaning "seven" and "Mau" meaning "three thousand six hundred square units"—which once spanned six of eight villages in Cam Thanh commune, bordered to the west and northwest by the estuary of the Thu Bon river. Since the eighteenth century, people brought Nipa palm from the Mekong region to plant in Cam Thanh commune. This species has proven its ability to thrive in this climatic region, reflecting both natural adaptation and human influence on the estuary landscape. The Cam Thanh Nipa wetland is a part of a mangrove ecosystem located in the lower reaches of Thu Bon river. It serve as a habitat for many aquatic species in the estuary, a seasonal refuge for migratory storks, and a natural barrier that break waves, wind, helping to protect ships from storms.

Together with the Tam Hai coral reef ecosystem, and Cham Island, the Cam Thanh Nipa wetland forms part of a marine-protected area network in Quang Nam (Huyen et al., 2015).

Cam Thanh Nipa wetland is not only a source of livelihood support for local people by providing food and raw materials for production, but also a rich mangrove ecosystem for aquatic animals in the estuary. Cam Thanh nipa wetland with many characteristics and functions of estuarine wetlands has provided a variety of goods and services to stakeholders (UNDP, 2012).

Due to the economic activities of local people and neighboring areas exploiting wetlands, the Cam Thanh nipa wetland area has gradually lost many wetland functions. As a result, wetland goods and services have declined, thereby reducing the income or living standards of people living on the exploitation of wetland resources.

Although there was no recorded data on the reduction of the forest area, most local residents in the community agree that during 1990s, the conversion of wetland into shrimp farms and salt-field—initially bringing subsantial profits to the—was one of the main reasons for the dramatic loss of forest cover in a short period. Road, bridge construction, and settlement have caused the scratching of nipa palm distribution which can be thought to be reduced in the fully ecological functions of ecosystem. Therefore, the cover of nipa palm was just around 65 ha, and concentrated only in village 2, village 3, village 7, and village 8 (The report of project: Reforestation and conservation nipa palm Forest for eco-tourism and sustainable development in Cam Thanh commune, 2009). Salt-making activities were currently disappeared because of its economic inefficiency. Shrimp farming is also proving to be economically inefficient. Instead, thatch cottage and souvenir farming created job and income for the villagers. The project of tourist service of homestay and other tourist activities formed the new economic activity in this area. Two later economic activities contributed to the expansions of the cover of nipa palm trees. Unfortunately, the increase in demand on goods and services of nipa palm wetland did not parallel the reforestation and conservation in the quantity and quality of nipa palm available for those uses (Fig. 5 and Table 3).

## 3.2 Data Collection

Reviewing the websites, published books and papers, expert opinion provided an overviews of wetland goods and services. Table 4 provides an overview of wetland goods and services defined by ecosystem processes and functions, as well as by users and nonusers.

After synthesizing the existing documents to identify the possible goods and services of estuarine wetlands, step 1 will be conducted. The main objectives of this step are (i) contribute to standardizing expert knowledge on wetlands and apply it to the specific case study of Cam Thanh nipa wetland, (ii) determine the perceptions of stakeholders and the benefits they receive from Cam Thanh Nipa forest, and (iii) determine the potential trade-offs between stakeholders in nipa wetlands. After

**Fig. 5** The cover of nipa palm forest. Source: Cam Thanh Commune People's Committee

**Table 3** Data on areas of wetland

| Nipa palm cover (ha) | Shrimp and alga pond (ha) | Natural surface water (ha) |
|---|---|---|
| 65 | 325 | 157 |

Sources: Author synthesized from Hoi An department of Science and Technology and Environment

completing step 1, the goods and services of Cam Thanh Nipa wetlands are formed, and stakeholders are also identified and understand the decision-making process of Cam Thanh nipa wetlands. Based on Turner et al.'s stakeholder classification, this study identifies stakeholders and then investigates stakeholders' perceptions about the benefits they receive from Cam Thanh nipa wetland, dividing them into internal stakeholders and external stakeholders. Internal stakeholders are those who live within the nipa wetland area, while external stakeholders are those who live outside the nipa wetland.

**Table 4** Goods and Services of Cam Thanh wetland

| Functions | Goods and Services | Values |
|---|---|---|
| ***Provisioning*** | | |
| Flora | Trees, reed, algae; crops (rice, maize, ...); medicinal plants | Direct use value |
| Fauna | Fish, birds, terrestrial animals; aquaculture; livestock; wild foods | Direct use value |
| Water | Drinking; washing; bathing | Direct use value |
| Materials | Timber; cotton; hemp; silk | Direct use value |
| Fuel | Wood fuel; minerals | Direct use value |
| ***Regulating*** | | |
| Climate regulation | Recreation and ecotourism | Indirect use value |
| Water flow regulation and control; | Flood control | Indirect use value |
| Water purification | Water supply for irrigation, production, and transportation | Indirect use value |
| Sediment stabilization | Aquaculture; Irrigation | Indirect use value |
| Soil formation | Erosion and prevention saline intrusion | Indirect use value |
| ***Cultural*** | | |
| History | Spiritual services | Indirect use value; Bequest value, Existence value |
| Education | Scientific research; Apprentice; School subject | Indirect use value |
| ***Supporting*** | | |
| Nutrient cycling; Biodiversity maintenance | Scientific research | Nonuse value |
| Wind absorb; Attenuates waves | Storm protection | Indirect use value |
| Sediment removal; Wildlife habitat | Recreation | Indirect use value |
| Seeding raising; breeding ground | Aquaculture; Fishing | Indirect use value |
| | | Option value |
| | | Existence value |
| | | Bequest value |

Source: Millennium Ecosystem Assessment (2013), Turner et al., (1997), Barbier (2018), Costanza et al., (2018), Lafoley et al., (2004), Finisdore et al., (2020)

The information necessary for this study was primarily obtained from stakeholders at various levels, including local, commune, city, provincial/national, and international organization. The required information was gathered using a combination of a variety of data collection methods including face-to-face interviews, group discussions, and field observations.

**Table 5** Selected villages, population and questionnaire distribution and expert and coordinator meeting

| | Villages | | | | | | | | |
|---|---|---|---|---|---|---|---|---|---|
| | 1 | 2 | 3 | 4 | 5 | 6 | 7 | 8 | Total |
| Household | 237 | 217 | 314 | 206 | 289 | 337 | 204 | 126 | 1930 |
| Population | 920 | 935 | 1131 | 861 | 1159 | 1296 | 891 | 483 | 7676 |
| Group discussion | 25 | 30 | 40 | 34 | 0 | 0 | 32 | 37 | 198 |
| Interviewed household | 23 | 67 | 76 | 21 | 0 | 0 | 67 | 55 | 309 |
| Sample | 22 | 63 | 69 | 19 | 0 | 0 | 58 | 50 | 281 |
| | **Expert and coordinator meeting** | | | | | | | | |
| Meeting with an officer from Cadastral and Environmental Unit, Commune | | | | | | | | | 1 |
| Meeting with an officer from Cadastral and Environmental Department, Hoi An city | | | | | | | | | 1 |
| Meeting with an officer from Fishery Department, Hoi An city | | | | | | | | | 1 |
| Meeting with a coordinator from GEF | | | | | | | | | 1 |
| Meeting with an expert from Cu Lao Cham Marine Protected Park Management Unit | | | | | | | | | 1 |

Source: The authors

The field study began with a reconnaissance visit to the area and meetings with the communal government to get first-hand information on the stakeholders, dominant economic activities, and location relevance to wetland. Subsequently, lists of stakeholders were composed and a strategic plan was developed to collect information from stakeholders (Table 5).

Data collection at the study site was carried out in six purposively selected villages (villages 1, 2, 3, 4, 7, 8) within the wetland area. These villages were chosen because they are the significant settlements suro unding the wetland, and their residents depend heavily on the goods and services provided by the wetland for their livelihoods. Moreover, each of the six villages includes professional groups that directly exploits natural resources from the wetland. Collectively, the selected villages offer a representative overview of the benifits people receive from the wetland ecosystem (Table 6).

For local group discussion, a meeting was held with communal officers to obtain the necessary permits for conducting surveys in Cam Thanh commune and to seek their assistance in contacting the head of villages and local residents. With the help of the communal officer and each village chief, people from six villages joined their own group meetings to discuss about the goods and services of the Nipa palm wetland (Fig. 6).

The information gathered from the group discussions was instrumental in developing the questionnaires used for subsequent face-to-face interviews. These group discussion focused on identifying various user groups and getting their views on the benefits as well as potential benefits they are receiving.

**Table 6** Villages and economic activity

| Village | Major economic activity |
|---|---|
| 1 and 2 | Offshore fishing  |
| 3 and 4 | Rice farming  |
| 5 and 6 | Shrimp and alga farming   |
| 7 | Thatched house   |
| 8 | Changing from finishing in the wetland to tourism   |

Source: The author

**Fig. 6** Meetings with local residents. Source: The author

After questionnaires for the survey were designed, a test survey was implemented to check:

- The terminology used in the questionnaire is appropriate for the local language
- Remove unnecessary question
- Time interview

A total of 309 questionnaires were completed using a random sampling technique with respondent households selected randomly. The number of respondents from each village was determined on a pro-rata basis. The survey was conducted from March to August, coinciding with the dry season which is the peak period for harvesting resources. Consequently, most households were working in the fields, making it challenging to find suitable time for interviews.

For external stakeholders, the assumption was that they might have a good knowledge of Nipa palm wetland because they have connected with this Nipa palm wetland for many reasons. The questions asked concerned the same topics as for local stakeholders: the definition of wetlands, their functions and users, good management of a wetland, and ways to manage them.

## 3.3 *Goods and Services of Cam Thanh Wetland*

Cam Thanh includes the area of residential land, agricultural land and water surface, in which the residential and farming land are private properties owned by households which are recorded in the estate book according to Vietnamese law. In contrast, the water surface is an open-access resource.

The villages in Cam Thanh are organized according to their primary economic activities. Village 1 and 2 consist of most households capturing fishery offshore in Nipa palm wetland. Village 3 and 4 are engaged in shrimp farming and alga farming. Village 5 and 6 are farming rice. Village 7 and 8 works with nipa palm exploitation to build thatch, souvenirs, and ecotourism.

From expert perspectives of Nipa palm wetland, previous study at the site and through the face-to-face interviews with the local residents (Tables 7 and 8).

**Table 7** Overview of the results of Cam Thanh wetland goods services

| Function | Goods and services identification | Stakeholders |
|---|---|---|
| Provisioning | • Leave and stalk | Local resident |
| | • Capture fishery | Local resident |
| | • Bird watching | Local resident |
| | • Alga farming | Local resident |
| | • Shrimp farming | Local resident |
| | • Wood fuel | Local resident |
| Regulating | • Erosion protection | Local resident |
| | • Recreational fishing and leisure | Local resident, national and international tourists |
| | • Flood control | Local resident |
| | • Water for shrimp and alga farming | Local resident |
| Culture | • Education | Local/city/provincial and national level |
| | • Historical relics | Local resident |
| Supporting | • Bird watching | Local and city level |
| | • Storm shelter for ships | Local and provincial level |
| | • Buffer of the world biosphere reserve | |
| | • Storm shelter for ships | Local resident |
| | • The costs to reforest nipa palm and to expand the wetland cover<br>• Willingness to pay of respondents to preserve the nipa palm for future generations<br>• The fund for nipa palm conservation | |

## 4 Conclusion and Discussion

Wetlands worldwide remain threatened despite numerous international, governmental, and local efforts to prevent or mitigate their. Market failures persist exist due to the public good nature of certain wetland goods and services and externalities from users upon other stakeholders, including nonusers.

This chapter identifies the goods and services provided by estuarine wetlands, and presents an exploratory case study of Cam Thanh. The approach used is stakeholder-based where stakeholders serve as key informants to access wetland goods and services. Consequently, some goods or services may be identified through scientific analysis but not be recognized by stakeholders, and thus their values are excluded from the wetland's list of goods and services.

The study identifies the quality of goods and services but does not quantify benefits. However, using knowledge to quantify all the goods and services at a site in a comparable way would provide valuable information for assessing the benefits they offer.

In environmental management, policymakers need to quantify the benefits of goods and services provided by a specific wetland to understand the impact of a development policies or human activities. However, due to limited time and

**Table 8** Identification of Cam Thanh wetland stakeholders

| | Local level | | | District | | | | Provincial/national | | | | International level | | | |
|---|---|---|---|---|---|---|---|---|---|---|---|---|---|---|---|
| Ecosystem goods and services | Rice Farmers | nipa palm crafter | Shrimp farmers | Officials | Others | Hoi An people | Cham Island resident | fishermen | Officials | Students, scientists | officials | NGOs | Tourists | Students, scientists | tourists |
| Leave and stem | | x | | x | | | | | | | | | | | |
| Fishery | x | x | x | x | x | | | | | | | | | | |
| Alga | | | x | x | | | | | | | | | | | |
| Leisure and recreation | x | x | x | x | x | x | | | x | | x | | x | | x |
| Bird watching | x | x | x | x | x | x | | | | | | | x | | x |
| Recreational fishing | x | x | x | x | x | x | | | | | | | x | | x |
| Education | | | | | | | | | | x | | | x | x | x |
| Hydrological regimes | x | x | x | x | x | | | | | | | | x | | |
| Historical relics | x | x | x | x | x | x | | | x | x | x | | x | x | x |
| Flood attenuation | x | x | x | x | x | x | | | x | | x | x | x | | |
| Biodiversity maintenance | x | x | x | x | x | | | | x | | x | x | x | | |
| Biological balance | x | x | x | x | x | x | | | x | | | | x | | |
| Climate regulation | x | x | x | x | x | x | | | x | | x | | x | | |

(continued)

**Table 8** (continued)

| Ecosystem goods and services | Local level | | | District | | | | Provincial/national | | | | International level | | | |
|---|---|---|---|---|---|---|---|---|---|---|---|---|---|---|---|
| | Rice Farmers | nipa palm crafter | Shrimp farmers | Officials | Others | Hoi An people | Cham Island resident | fishermen | Officials | Students, scientists | officials | NGOs | Tourists | Students, scientists | tourists |
| Storm shelter for ships | | | | | | | | x | x | | x | | x | | |
| Buffer of the world biosphere reserve | | | | | | | x | | x | | x | | x | | |
| Silt filter and waste barrier | | | | | | | x | | x | | x | | x | | |
| Erosion protection | x | x | x | x | x | | | | x | | x | | x | | |
| Wind and waves absorbability | x | x | x | x | x | | | | x | | x | | x | | |
| Wastewater purification | x | x | x | x | x | | | | x | | x | | x | | |

Source: The author

resources, conducting a full quantification is often challenging. Therefore, a qualitative assessment based on the perceptions of involved stakeholders offers valuable information to support environmental policy and management decisions. Defining the ecosystem in terms of goods and services through stakeholder involvement also makes the concept more accessible and understandable to policymakers and nonscientists.

Although the results of defining goods and services of wetlands come from a specific case study, the assessment may not capture all the benefits wetlands provide to humans and the environment. Therefore, further research is needed to identify additional benefits, which will later assist in quantifying the values of these goods and services. The less careful assessment of the benefits or values of open natural resources' goods and services, the greater the risk of overexploitation of wetlands. Despite the development of more valuation methods, benefits that are not marketed generally remain more difficult to quantify.

**Proposed solutions for sustainable management and conservation of Cam Thanh nipa wetland**

Solutions for the conservation and reforestation of nipa wetlands should consider stakeholders' interests, reflecting people's awareness and concern for it. Local-level solutions should consider the interests of community members living within the nipa wetland area, where people's daily lives and livelihoods largely depend on the goods and services of the wetlands. Solutions for stakeholders living outside the wetlands should focus on their concerns with the characteristics and functions of nipa wetlands, resolving conflicts of interest between groups with opposing interests toward the conservation and development of nipa wetlands on a larger scale.

For people who are exploiting coconut leaves and sheaths to build huts and souvenirs, there should be technical solutions to guide farmers to cut leaves in the right season and the right amount, ensuring that the remaining leaves have enough time to grow and sprout branches in time for the next harvest. Monitor and have effective sanctions to prevent the cutting and overexploitation of coconut leaves.

For people who are catching shrimp, fish, crabs, and aquatic creatures living in nipa palm wetlands for food purposes and selling at the market, there should be guidance on sustainable fishing techniques, not allowing overfishing, and at the same time preventing people from neighboring areas from illegally and unsustainably exploiting this resource. For farmers who have shrimp ponds that are currently abandoned due to pollution, consider converting these shrimp ponds into nipa palm plantations, converting the land use while also ensuring the benefits they can receive from this conversion, such as receiving financial support from tourism companies that are using nipa palms for ecotourism.

For those using nipa palm wetlands for ecotourism, the current development plan involves converting some wetland areas into ecotourism zones. This local tourism plan aims to improve residents' livelihoods while reversing wetland reduction and fragmentation caused by agricultural activities. As the conversion progresses, it's essential to consider the benefits for farmers shifting from coconut and shrimp farming to tourism-related occupations, including financial support from tourism revenues. The impact of tourism on the gradual encroachment of nipa wetlands toward the sea depends on the income generated from these activities. This conservation effort not only expands the wetland area but also enhances its ecological functions.

For the entire community living in the nipa wetland area, raising awareness and knowledge about the value of wetland goods and services is essential. This helps them understand both the benefits these wetlands bring to society and the benefits they receive from the conservation and development of nipa wetland. As a result, they will participate fully, actively and effectively in managing, planting, and nurturing nipa trees in their locality.

For households currently encroaching on wetlands to convert them into residential areas, stricter laws must be enforced to prevent such this encroachment. Illegal, encroachers should be required to restore the wetlands to their original state and manage their nipa areas more sustainably to ensure long-term livelihoods.

## References

Barbier EB (2018) The value of Coastal Wetland Ecosystem Services. Elsevier B.V

Costanza R et al (2018) The value of the world's ecosystem services and natural capital. Ecol Econ 25(1):3–15

Cowardin LM et al (1979) Classification of wetlands and deepwater habitats of the United States. US Fish and Wildlife Service, Washington, DC

Finisdore J et al (2020) The 18 benefits of using ecosystem services classification systems. Ecosystem Services, 45

Grimble R, Wellard K (1997) Stakeholder methodologies in natural resource management: a review of principles, contexts, experiences and opportunities. Agric Syst 55(2):173–193

Helliwell DR (1969). Valuation of wildlife resources. Regional Studies, 3(1), 41–47

Huyen N et al (2015) Marine biodiversity and wetlands conservation in Vietnam. Ha Noi

Lafoley D et al (2004) The ecosystem approach: coherent actions for marine and coastal environments. English 65

Millennium Ecosystem Assessment (2013) Ecosystems and human well-being

Murat E, Ahmed A (2023) Introductory chapter: wetlands – characteristics, functions, and values. Wetlands - New Perspective. IntechOpen. https://doi.org/10.5772/intechopen.111469

Rai S, Raleng A (2011) Ecological studies of wetland ecosystem in manipur valley from management perspective (Vol. Ecosystems biodiversity). In Tech EBooks. https://doi.org/10.5772/24093

Springgay E, Janzen S (2019) The forest-water nexus: an international perspective forests 10(10)

Turner R et al. (1997) Ecological economic analysis of wetland and scocial science intergration 21(3): 295–316

Turner RK et al (2000) Ecological-economic analysis of wetlands: scientific integration for management and policy. Ecol Econ 35(1):7–23

UNDP (2012) Rehabilitation and conservation of Nyapalms at Cam Thanh Commune, Hoi An city, Quang Nam. Global Environment Facility, Ha Noi

WHO (2005) Health service planning and policy-making: a toolkit for nurses and midwives (Vol. Module 2: Stakeholder analysis and networks). WHO Regional Office for the Western Pacific, Regional Office for the Western Pacific: World Health Organization. Retrieved from https://iris.who.int/handle/10665/207061

**Trang-Huyen Le** is a lecturer and researcher in environmental economics, and microeconomics at Foreign Trade University. Her research expertise focuses on environmental economics and management, environment and trade, environmental policy, and environmental impact assessment. She has published in journals such as *Natural Resources and Environment Journal* and *Journal of International Economics and Management.*

**Assoc. Prof. Thuy Anh-Tu** (Ph.D. in Economics, Iowa State University, US) is Editor-in-Chief of the *Journal of International Economics and Management* (Foreign Trade University). Her research expertise focuses on international economics, environmental economics, and labor economics. She has published in international peer-reviewed journals such as *Ecological Economics* and *Palgrave Macmillan Asian Business Series.*

# Pathway to Sustainable Development for Vietnam: Mitigating Greenhouse Gas Emissions in Transportation Activities

**Tuan-Anh Bui, Huong-Thu Le, and Hanh-Thi Vu**

**Abstract** Climate change has been one of the most pressing and controversial issues that have drawn the attention of global academic researchers, policymakers, and international donors. It is widely recognized as a consequence of greenhouse gas (GHG) emissions. In Vietnam, the transportation sector is a major driver of GHG emissions, accounting for approximately 20% of all carbon emissions. This chapter investigated different aspects of Vietnam's transportation that are the main contributors to a high-carbon transport system: infrastructure, vehicle types, and their fuel consumption. Accordingly, Vietnam is considered among the countries most vulnerable to climate change in terms of environmental degradation, public health risks, and socioeconomic impacts. In addition to assessing mitigation strategies adopted by Vietnam's public and private sectors, the study proposed recommendations to promote climate-smart transport through three major shifts: fuel shift, modal shift, and energy efficiency improvement.

## 1 Introduction

In September 2015, the United Nations Member States collectively endorsed the 2030 Agenda for Sustainable Development, establishing a unified framework aimed at fostering peace, prosperity, and well-being for both humanity and the planet. Out

T.-A. Bui
Faculty of Business Administration, Foreign Trade University, Dong Da, Ha Noi, Vietnam
e-mail: buianhtuan.bgh@ftu.edu.vn

H.-T. Le
FPT Software Company Limited, Cau Giay, Hanoi, Vietnam
e-mail: huonglt26@fpt.com

H.-T. Vu (✉)
Center for Foreign Economic Research Training and Consultancy, Foreign Trade University, Dong Da, Ha Noi, Vietnam
e-mail: hanhvt@ftu.edu.vn

 
H.-T. Vu, L.-P. Dana (eds.), *Vietnam on the Road to Sustainable Economic, Social and Business Development*, Frontiers in South and Southeast Asian Development Research, https://doi.org/10.1007/978-981-96-4371-4_4

of 17 goals to transform the world, goal 13, "Climate Action," is an urgent call for taking actions to combat climate change and prevent its long-term effects on the environment, human society, and economy (Espagne, 2021). Global warming, ocean acidification, sea level rising, and other consequences of climate change are having significant adverse effects on coastal regions and low-lying coastal nations. While developed countries are estimated to bear 20–25% of the global impact of damages, developing countries will bear 75–80% of it, which costs up to USD175 billion annually by 2030 for mitigation (World Bank, 2010).

The primary cause of global warming and climate change is greenhouse gas emissions. There is a steady growth of the annual GHG emissions worldwide during the past years, reaching 53.0 Gt CO2eq in 2023 (Tiseo, 2024). Out of the top 10 emitters, China leads with 30.1% of worldwide emissions, followed by the United States at 11.3%, the European Union 27 at 6.1%, and India at 7.8%. These countries, characterized by substantial populations and economies, collectively represent half of the global population and nearly 62.3% of the world's GDP (World Bank, 2023).

As being among the underdeveloped nations in the mid-1980s, Vietnam has rapidly progressed its socioeconomy in a remarkable and impressive way, becoming a middle-income country since 2010 (World Bank, 2013). Vietnam recorded a high economic growth rate at 7.5% in 2018. Between 1990 and 2018, Vietnam's GDP per capita grew 27-fold, while the national poverty headcount ratio decreased significantly, dropping from 58% in the early 1990s to just 6.7% by 2018. However, Vietnam is not an exception in the process of industrialization, following the same path as other developing countries like China, India. These reforms were accompanied by environmental pollution. The carbon dioxide emissions have soared nearly seven times from 0.29 (1990) to 3.31 metric tons per capita (2022) (World Bank, 2022), which makes Vietnam the 13th most carbon-intensive economy globally and also one of the most climate-vulnerable countries. Therefore, addressing this global concern has become a national priority in order to balance socioeconomic development with environmental stability. By 2030, Vietnam has committed to reduce greenhouse gas emissions by 8% across various sectors such as energy, agriculture, waste, LULUCF, and industrial processes (World Bank, 2019).

Transportation is the subsector of energy that makes a substantial contribution to global GHG emissions, accounting for 24% of the global carbon emissions in 2030 (International Energy Agency 2024). Particularly in Vietnam, the transport sector emitted about 30.6 million tons CO2e in 2014, accounting for 18% of total CO2e emissions and are projected to increase by 7% annually, reaching nearly 90 million tons of CO2 emissions by 2030 (World Bank, 2019). Although serious initiatives by the Government and Ministry of Transport to reduce CO2e in road transportation by covering vehicles using internal combustion engines to electric vehicles and other types of vehicles using green energy, the contribution of Vietnam 's transport emissions did not basically decrease (SLOCAT, 2023). Transport activities refer to all types and modes of transportation. In terms of purposes, it can be categorized into two types: passenger transportation and freight transportation. In terms of forms, it can be categorized into four modes: road, railway, airway, and waterway transport.

The excessive use of nonrenewable energy fossil fuels in the transportation service results in high carbon emission output, posing detrimental impacts on the environment.

Despite the significantly high contribution of carbon emissions from transport activities, carbon management in transportation in Vietnam seems to be under-researched. This chapter aims to provide an overview of the transportation sector and the associated greenhouse gas emissions, as well as to propose a vision and strategy for climate-smart transport operations to minimize the carbon footprint.

The analytical research is organized as follows. The next section provides a brief overview of carbon emissions in the transportation sector, identifying issues and its impacts on economic, social, and environmental aspects. This is followed by the assessment of mitigation policies and strategies adopted by the government and private sector. The fourth and fifth sections offer several carbon management lessons learnt from China and give recommendations for Vietnam. We hope that these findings can provide actionable insights and recommendations, contributing to a Vietnam's low-carbon and resilient transportation system.

## 2 An Overview of GHG Emissions from Transport Sector in Vietnam

### *2.1 The Current Situation of Vietnam's Transport Sector*

Since the political and economic reforms known as Doi Moi in 1986, Vietnam has witnessed a socioeconomic development success story. Its transformation from a low-to a middle-income country since 2010 has been reflected in the remarkable rise in gross domestic product (GDP) per capita, from USD239 in 1985 to USD3,817.20 in 2023. As a result of economic growth, the demand for transportation is expanding rapidly. It is considered one of the fastest-growing industries in Vietnam with significant development in infrastructure and a sharp increase in traffic flow. In 2023, the transportation and warehousing sector contributed 4.92% to the country's GDP.

#### 2.1.1 Transport Infrastructure

Road Transport

Vietnam has spent tremendous effort in developing its infrastructure in order to meet the growing needs of transportation. By 2022, the total length of Vietnam's road system is about 595,201 km, of which 25,560 km comprises national roads (including national highways and expressways) ways only account for 4.45%. National highways are regarded as arteries of the entire road network, creating the

North-South and East-West corridors. In the North, the national highway network is designed to shape as a hub-and-spoke model with its center being Hanoi capital, while in the south, it forms a grid. In general, the current highway system covers the whole national territory and plays a pivotal role in connecting key population centers with industrial zones, economic zones, airports, and seaports (Table 1).

The entire road network is well developed but still in poor quality. Only 9.1% of the network (51,258 km) is paved. Most of the roads are narrow and incompatible with the traffic flow. The proportion of the national highways (main roads) is quite similar with that of provincial highways (secondary roads), while in developed countries, the total length of secondary roads should be at least twice that of main roads. It has become a serious illness that weakens many "bodies–societies" that have not taken anything well of Vietnam (Nguyen Minh Quyen et al., 2019).

Railway

Vietnam's current railway network of 3,143 km comprises 15 main lines (2,531 km) and several branch lines (612 km), running through 34 cities and provinces nationwide. The primary railway line is the Hanoi–Ho Chi Minh City route (also known as Thong Nhat route) with 1,726 km long. Other lines are concentrated in the north with Hanoi being a central hub, such as Hanoi–Hai Phong route (102 km) and Hanoi–Quan Trieu (75 km). As outlined in the national railway network plan, the Lao Cai–Hanoi–Hai Phong railway line will cover approximately 390.9 km, featuring double-track, with a track gauge of 1,435 mm. This line is intended to connect to Kunming in Yunnan Province, China. The Ministry of Transport is currently preparing the pre-feasibility report for the project, which has an estimated total investment of $11 billion (VCCI, 2024). The current railway infrastructure is underdeveloped and poorly maintained since most of it was constructed 20–30 years ago. In addition, the railway system consists of 297 stations, mostly in small scale, equipped with only few tracks and in short length (many stations are less than 325 m in length), making it unfavorable for transport operation.

**Table 1** Vietnam's road network in 2022

| Road class | Length (km) | Proportion (%) |
|---|---|---|
| Expressway | 1850 | 0.31 |
| National Highway | 25,560 | 4.29 |
| Provincial Highway | 25,741 | 4.32 |
| Local (district/commune) roads | 216,135 | 36.31 |
| Urban roads | 27,688 | 4.65 |
| Other rural roads | 247,082 | 41.51 |
| Specialized roads | 51,145 | 8.6 |
| Total | 595,201 | 100 |

Source: Compilation by the authors

## Waterway

### *Inland Waterway*

Vietnam has 2,360 rivers and canals with a total length of 220,000 km. Out of the network, only 19% (42,000 km) is considered navigable and 12% (26,650.8 km) has been placed under management and operation by central and local governments (Table 2).

The Northern River network has 17 main routes with a total length of 4,500 km. It can be defined by four main river systems: the Red River, Thai Binh River, Duong River, and Luoc River. The channels have widths ranging from 30 to 37 m and depths between 1.5 and 3.6 m. The region's river conditions are influenced by northern hydrometeorological patterns, with a rainy season from May to October and a dry season from November to May. The water level fluctuates at 5–7 m between these two seasons.

The Central region has 10 routes flowing independently west to east (from mountains to sea) across 19 provinces and cities from Thanh Hoa to Binh Thuan without forming a network. Because of steep terrain and depth limitations, river navigation is generally unviable except for sections from National Highway 1 to river estuaries.

The Southern region has 18 routes with a total length of 6,500 km, formed by the Dong Nai and Mekong River systems. Water transport on these rivers is more accessible than in the North, with channel widths ranging from 30 to 100 m and depths of 2.5 to 4 m.

Maritime

With 44 seaports and 229 terminals, Vietnam has 3,400 km of coastline along one of the world's busiest sea cargo lanes. Thanks to its favorable geographical location, Vietnam serves as a transshipment hub for imported goods from Asian countries (via Cat Lai port) and exported goods to the United States and European Union (via Cai Mep–Thi Vai port). Seaports in Vietnam are classified into four categories based on their operational purposes (Table 3):

While railways and dedicated expressways are used for connecting major ports nationwide in developed countries, Vietnam's seaport system has exclusively Hai Phong port connected to the railway despite low efficiency (Cai Lan port has

**Table 2** Scale of inland waterway in Vietnam

<table>
<tr><td colspan="4">Vietnam inland waterways: 220,000 km</td></tr>
<tr><td colspan="3">Navigational length: 42,000 km (19%)</td><td rowspan="6">Non-navigational length: 178,000 km (81%)</td></tr>
<tr><td colspan="2">Under management: 26,650.8 km (63%)</td><td rowspan="5">Not under management: 15,349.2 km (37%)</td></tr>
<tr><td>By central government: 7,071.8 km</td><td>By local government: 19,579 km</td></tr>
<tr><td colspan="2">Of which major Northern routes: 4,500 km</td></tr>
<tr><td colspan="2">Of which major Southern routes: 6,500 km</td></tr>
<tr><td colspan="2">Of which nonmajor routes: 31,000 km</td></tr>
</table>

Source: MOIT (2019)

**Table 3** Vietnam's seaports classifications

| Class | Number of ports | Operational purposes |
|---|---|---|
| IA | 2 | Large-scale seaports to serve the socioeconomic development of the whole country or of the inter-region and serve as an international transshipment point |
| I | 11 | Large-scale seaports to serve the socioeconomic development of the whole country or of the inter-region |
| II | 7 | Medium-scale seaports to serve the socioeconomic development of localities |
| III | 14 | Dedicated seaports close to oil derrick at sea to export crude oil |

Source: Decision 804/QĐ-TTg (2022)

received investments but remains unable to operate due to lack of synchronous gauge) and no expressway dedicated for freight transportation. In addition, almost all of Vietnam's ports have limited wharf length (450–900 m), which makes container ships queue for 6–8 h for unloading. These disadvantages lower the operational efficiency of transport in terms of time and cost.

Airway

Currently, there are 23 airports in operation nationwide, including:

- 11 international airports: Noi Bai, Cat Bi, Da Nang, Cam Ranh, Phu Bai, Tan Son Nhat, Can Tho, Phu Quoc, Van Don, Vinh, Phu Cat.
- 12 domestic airports: Dien Bien Phu, Dong Hoi, Vinh, Phu Cat, Tuy Hoa, Pleiku, Chu Lai, Buon Ma Thuot, Ca Mau, Rach Gia, Con Dao, Tho Xuan.

Many airports in Vietnam have reached their full capacity and are overloaded. For example, in 2023, Tan Son Nhat airport welcomed 41 million passengers, 1.4 times as high as its designed capacity and Noi Bai airport welcomed up to 29 million passengers, 1.2 times as high as its designed capacity. Out of all civil airports nationwide, only 4 airports have separate cargo-handling terminals. For the other airports, all cargo is handled in the passenger terminals (Table 4).

### 2.1.2 Traffic Flow (Table 5)

The development in infrastructure reflects a rapid increase in mobility across both pasenger and freight traffic. In the 2019–2022 period, the annual growth rates in passenger and freight traffic strongly fluctuated. Prior Covid-19 pandemic, the passenger traffic experienced an average annual growth of approximately 10%, while freight traffic grew by 7%. However, the pandemic caused a significant downturn in 2020, with passenger traffic plummeting by 32.8% and freight traffic declining by 2.9%. In 2022, the sector demonstrated a strong recovery, with the passenger and

**Table 4** Capacity of major airports that have cargo-handling terminals in 2023

| No. | Airport name | Location | Capacity for passenger terminal (million people/year) | Capacity for cargo-handling terminal (ton/year) |
|---|---|---|---|---|
| 1 | Noi Bai | Ha Noi | 29 | 700,000 |
| 2 | Tan Son Nhat | Ho Chi Minh | 41 | 700,000 |
| 3 | Da Nang | Da Nang | 6.3 | 20,000 |
| 4 | Cam Ranh | Khanh Hoa | 5.7 | 10,000 |

Source: Compilation by the authors

**Table 5** Volume of passengers/freight traffic by types of transport

| | Transport mode | 2017 | 2018 | 2019 | 2020 | 2021 | Prel. 2022 |
|---|---|---|---|---|---|---|---|
| Passenger (billion passengers-km) | Roadway | 125.4 | 130.8 | 114.1 | 112.7 | 78.0 | 123.5 |
| | Railway | 3.6 | 3.5 | 3.1 | 1.5 | 0.6 | 1.6 |
| | Waterway | 3.5 | 4.5 | 6.1 | 2.7 | 1.2 | 3.5 |
| | Airway | 54.3 | 67.9 | 77.4 | 34.1 | 13.9 | 54.9 |
| | Total | 186.8 | 206.7 | 230.7 | 151.1 | 93.8 | 183.6 |
| Freight (billion tons-km) | Roadway | 63.5 | 71.0 | 76.5 | 73.5 | 75.3 | 89.9 |
| | Railway | 3.6 | 4.0 | 3.7 | 3.8 | 4.1 | 4.5 |
| | Inland waterway | 47.8 | 52.6 | 56.0 | 52.9 | 24.8 | 31.6 |
| | Maritime | 140.3 | 144.6 | 154.8 | 152.6 | 70.1 | 91.2 |
| | Airway | 5.0 | 5.6 | 6.2 | 3.6 | 14.1 | 6.1 |
| | Total | 260.2 | 277.9 | 297.2 | 286.3 | 188.3 | 223.4 |

Source: General Statistics Office of Vietnam (2024)

freight traffic increased by 95.7% and 18.6%, respectively. Such rapid growth in mobility has contributed to significant economic growth and poverty reduction in Vietnam on one hand, resulted in negative environmental consequences, one of which is high greenhouse gas emissions on the other hand.

## 2.2 *The Issue of GHG Emissions from Transport Sector in Vietnam*

The transport sector has become a significant driver of greenhouse gases (GHG), emitting around 64.3 million tons CO2 in 2025, equivalent to 18% of Vietnam's total GHG emissions (Nguyen, 2024). Emissions per passenger or ton of freight volume can vary significantly depending on different modes of transport with respect to quantity and quality of the vehicles and its fuel consumption. There are four types of fuels that are used for transportation: gasoline is widely consumed by road vehicles, such as motorcycles, passenger cars, light duty vehicles; diesel is used in road, railway, and inland waterway transport; fuel oil is only used for maritime vehicles; kerosene is only used in aviation.

Table 6 Vietnam's truck fleet characteristics

| Truck type | Truck age (years) | Market size (%) | Fuel efficiency (L/100 km) |
|---|---|---|---|
| 0–5 T | 7.68 | 68 | 11 |
| 5–10 T | 5.91 | 11 | 18 |
| 10–20 T | 8.83 | 14 | 28 |
| >20 T | 5.82 | 7 | 34 |

Source: World Bank (2019)

### 2.2.1 Roadway

Among four types of transport modes, road transport contributed to 90% of transport-related emissions. One of the reasons that made road transport the largest emitter was that it carried 94% of passengers and 76% of freight tons.

For passenger transport, the high volume of private vehicles is the biggest source of transport emissions and the major reason behind traffic jams in Vietnam. While public vehicles as well as nonmotorized transport are preferred for sustainability, private vehicles dominate transport practices in Vietnam. Specifically, motorcycles make up for 94% of the total road fleet, followed by passenger cars. Public passenger transport (bus, coach) and nonmotorized vehicles (bicycles, electric vehicles) account for a minimal proportion of the overall mode share. Besides, over one-fourth of the motorcycles in Vietnam are more than 10 years of age. Old motor vehicles are proven to emit much greater levels of toxic gases than those that are maintained periodically.

For freight transport, Vietnam's truck fleet is regarded as unfriendly to the environment due to low fuel efficiency. Truck fuel consumption depends on the size and age of the truck; to be more specific, the bigger the vehicle, the higher the emissions are released; the older the engine, the less fuel efficient it is (Table 6).

In Vietnam, the 0–10 tonnage trucks are widely used for domestic shipping, accounting for the vast majority of the market share (79%). However, the large number of low tonnage trucks carry the same freight volume on the same intercity routes adversely leads to unnecessary GHG emissions. In contrast, while high-capacity trucks consume more fuel and generate more emissions, they exhibit better fuel efficiency because of the increase of volume of freight traffic (volume of freight through a length of transportation distance). Moreover, nearly 95% of the national truck fleet is more than 5 years of age, with 31% of that is older than 31 years (Lamet al., 2019). In addition, empty backhauls and less than truckload (LTL) are one of the logistics operations challenges to Vietnam. The fact that the trucks operate on road with no or suboptimal loads results in increase in both logistics cost and greenhouse gas emissions.

### 2.2.2 Railway

Rail transport in Vietnam represents a small and declining portion of both passenger and freight volumes, primarily due to the poor condition of the infrastructure, which offers slow speeds and low-capacity services. In general, the main train connecting Ho Chi Minh and Hanoi is usually operated at an average speed of 50–60 km per hour, while the average speed of interprovincial trains is 30–40 km per hour. The current system comprises only 1-m-wide single track, but if it was upgraded into double-track railway, average speed could increase up to 120 km per hour. The low average speed results in low fuel efficiency.

### 2.2.3 Waterway

While waterway is not a popular mode for passenger transport, it accounted for greatest share of freight flows in Vietnam thanks to its advantage of the travel distance over roadway. Same as roadway transport, the emissions from waterways and maritime transport are also influenced by the size and age of vessels deployed. In Vietnam, bulk cargo is the primary commodity transported via inland waterways. It is typically transported by small vessels with a deadweight tonnage (DWT) of less than 250, and sometimes even less than 100 DWT. Meanwhile, the average cargo vessel fleet in Western Europe has a carrying capacity of 2500 DWT. This discrepancy results in higher fuel consumption per ton of freight traffic in Vietnam, with emissions ranging from 30 to 160 gCO2e. The fuel efficiency is proven to increase if vessels larger than 250 DWT are put into use. At the same time, most of Vietnam's fleet is technically old and outdated (usually more than 15 years of age). Low combustion efficiency and no installed scrubbers results in high levels of SO2, CO2, CO, NO2 emissions.

### 2.2.4 Airway

Aviation sector is a rapidly expanding sector in Vietnam and also the least polluting mode of transportation. The number of passenger kilometers traveled by domestic civil aviation grew significantly from 4.4 billion PKT in 2000 to 11.0 billion PKT in 2014, and it is projected to reach 36.5 billion PKT by 2030. The cruise emissions are dependent on the length of the flight. Specifically, an aircraft's emissions are higher during the take-off and landing phases, so the longer the flight is, the more kilometers the plane will have to soften the impact of these two phases.

## 2.3 *GHG Emissions' Impacts on Vietnam's Environment, Society, and Economy*

### 2.3.1 Impacts on the Environment

Due to its carbon intensity, Vietnam is among the most climate-vulnerable countries in terms of increasing sea levels, higher temperatures, and chronic and extreme weather events.

Under a low GHG concentration scenario, Vietnam's average temperature is expected to rise by 0.6–0.8 °C by the early twenty-first century, 1.3–1.7 °C by mid-century, and 1.9–2.4 °C in the North and 1.7–1.9 °C in the South by the end of the century. The average sea level rise along Vietnam's coast is expected to be 53 cm (ranging from 32 cm to 76 cm) by 2100 (World Bank, 2021). Under a high GHG concentration scenario, the nationwide average temperature is projected to be 0.8–1.1 °C at the start of the century, 1.8–2.3 °C by mid-century, and 3.3–4.0 °C in the North and 3.0–3.5 °C in the South by the century's end. The sea level rise in this scenario could reach 73 cm (ranging from 49 cm to 103 cm) by 2100.

With a long coastline facing the East Sea, the country tends to be exposed to tropical storms and typhoons more frequently. While the number of storms and tropical depressions may remain relatively consistent, they tend to be more concentrated toward the end of the storm season, which also coincides with a peak time for storms in the South. While the Mekong Delta is facing the threats from saltwater intrusion, Northern Mountain and Central areas are prone to landslides and flash flooding due to intense rainfall events. At the same time, the country would face greater risks and vulnerabilities to drought and water shortages, leading to increasing desertification in Vietnam. Although the frequency of severe cold weather days is decreasing, cold waves have become more unpredictable, with their intensity and occurrence varying from year to year. In 2017, Vietnam experienced most of extreme weather events, except for tsunamis. Over the past 25 years, natural disasters have caused 1$ of GDP loss on average (World Bank, 2019).

### 2.3.2 Impacts on National Public Health

The detrimental effects of air pollution on public health are growing concerns in many emerging countries. Fine particles emitted from vehicles into the air can penetrate deeply into the lungs and cardiovascular systems, leading to conditions such as heart disease, stroke, chronic obstructive pulmonary disease, lung cancer, and pneumonia. In Vietnam, cardiovascular disease is the reason for 33% deaths, followed by cancers and chronic respiratory diseases at 19% and 6%, respectively (Vietnamnet, 2024). In addition to air pollution-related diseases, the impacts of climate change, including higher temperatures and shifts in rainfall patterns, are expected to contribute to more frequent occurrence of infectious diseases. Malaria and dengue fever are highly sensitive to weather conditions with respect to temperature, rainfall, and humidity, which means only a minor sign of climate change can accelerate its transmission potential rapidly (Wang et al., 2023).

### 2.3.3 Socio Economic Impacts

The impacts of climate change, including such as rising sea levels, heavy floods, saltwater intrusion, droughts, and extreme weather events, are increasingly visible, taking a toll on Vietnam's socioeconomic development.

It is projected to damage 3.2% of Vietnam's GDP each year, especially in the agricultural and manufacturing sector. Firstly, for the agricultural sector, the cultivation land will be narrowed down, resulting in lower crop yields and reduced livestock production. Vietnam's current agricultural land spans approximately 28 million ha. If the sea level rises by 1 meter, about 40% of Mekong Delta, 11% of Red Delta and 3% of other coastal provinces will be inundated. By 2100, the Mekong Delta region could lose 89,473 ha of land, resulting in a loss of around 7.6 million tons of rice annually if sea levels rise by 100 cm (DCC, 2021). At the same time, saline intrusion will affect 45% of the lowland in the Mekong Delta. The combination of sea level rise and increased salinity will reduce grazing areas, damage breeding facilities, and degrade drinking water quality for livestock. Further, changes in rainfall patterns and temperatures will also disrupt irrigation, negatively affecting crop yields and distribution. For example, rice yield of spring crop is predicted to decrease by 16.5% by 2070, while the winter crop will see a smaller decrease of 5%. In contrast, winter maize yields may increase in the Northern Delta but decrease in the Central and South regions. Decline in rice productivity will deteriorate food security at both national and global level since Vietnam is considered a leading agricultural exporter in the world.

In the fishery and aquaculture sector, approximately 480,000 people in Vietnam are directly engaged in fishing activities, with an additional 100,000 working in seafood processing and around 2,140,000 involved in related services. In 10 provinces spanning from Quang Ninh to Thua Thien—Hue in the North and North Central regions—typhoons are projected to result in annual losses ranging from VND 60 billion to VND 115 billion while rising temperatures could cause losses of approximately VND 410 billion and VND 445 billion by 2050.

For the industry sector, if Vietnam experiences a 1-m rise in sea level, 20 industrial provinces would be inundated. Most provinces with many manufacturing firms affected are located in the Red Delta and Mekong Delta area, which contribute 56% of national industrial production by value. Ho Chi Minh City alone has 16 industrial zones that would be at high risk of inundation and storm surges, while the overall figure for the southeast region would be 55 estates (ADB, 2013). Raw materials, especially for food processing, garment and textile industries, will significantly decline because of the supply chain disruptions in the Mekong Delta provinces, which was the most vulnerable area in Vietnam. This puts pressure on the national economic restructuring from the manufacturing industry to the service sector economic base.

Due to the narrowing of residential areas and cultivation land, coastal communities would face the crisis of land loss and job loss. It is estimated that a 1-m sea-level rise could submerge up to 39% of the Mekong Delta, impacting 35% of the population and jeopardizing 40.5% of the region's total rice production (ADB, 2013). As a result, climate-induced migration is projected to occur. Many migrants concentrate on metropolitan cities, namely Hai Phong, Hanoi, Da Nang, and Ho Chi Minh, or in industrial zones such as Bac Ninh, Thai Nguyen, and Binh Duong. Settlements of climate refugees on this big scale would require special attention from the government to respond to emerging health, economic, and safety concerns.

A sea level rise of 100 cm poses a significant threat to Vietnam's transport network, with projections indicating that 12% of provincial roadways and 4% of railways could be submerged. The total length of road infrastructure permanently submerged could reach 11,000 km, including almost 90% in the Mekong Delta region. In the worst-case scenario, the total costs for new construction, maintenance, and repairs nationwide could reach up to US$55 billion by 2050 (Chinowsky et al., 2015). In addition, sea rise and saltwater intrusions take a toll on the foundation of coastal transport infrastructure, including seaports and railways, causing a decline in traffic works' life expectancy. An unsustainable transportation system could disrupt the supply chain and reduce the efficiency of logistics activities.

# 3 Vietnam Effort in Reducing GHG Emissions Caused by Transportation Activities

## *3.1 Responses from Government*

### 3.1.1 Policies of Climate Change Mitigation and Adaptation

Vietnam ratified several international conventions, namely United Nations Framework Convention on Climate Change (UNFCC) in 1994, the Kyoto Protocol in 2002, and the Doha Amendment in 2015. In alignment with the objectives of the UNFCCC, Vietnam submitted its Intended Nationally Determined Contribution (INDC) in September 2015. This submission was later officially adopted as the country's Nationally Determined Contribution (NDC) to the UNFCCC. Accordingly, several policies for climate change mitigation and adaptation have been adopted. Most of them are only recently catching up, with a focus on law enforcement and strategic frameworks for emission reduction and infrastructure development (Table 7).

The mitigation-related legal documents include the Law on Economical and Efficient Use of Energy (2011) and the Law on Environmental Protection (2014, 2020). Chapter IV of the Law on Economical and Efficient Use of Energy outlines measures to promote energy efficiency in the transport sector. These measures include optimizing transportation routes and vehicles, setting regulations for vehicle maintenance and repair, and implementing technological solutions for both

**Table 7** Targets for the GHG reduction task of the green growth strategy

| Period | Target | Related framework |
|---|---|---|
| 2011–2020 | • Reducing greenhouse gas emissions intensity by 8–10% compared to 2010<br>• Reducing energy consumption on GDP by 1–1.5% per year<br>• Reducing greenhouse gas emissions in the energy activity from 10% to 20% compared with the normal development plan | Decision 403/QĐ-TTg 2014: Approval of the national Action Plan on Green Growth in Viet Nam for the period of 2014–2020<br>Decision No. 1670/QD-TTg 2017: Approval for the Target Program for climate change adaptation and green growth for the period of 2016–2020 |
| 2030 | • Reducing greenhouse gas emissions intensity per unit of GDP at least 15% compared to 2014<br>• Reducing greenhouse gas emissions in the energy consumption per unit of GDP by 1.0-1.5% on average for the 2021–2030 | Decision No. 1658/QD-TTg 2021 To ratify the Green Growth Strategy for the 2021–2030 period, vision towards 2050 |
| 2050 | • Reduce GHG emissions by 1.5–2% annually | Decision No. 1055/QD-TTg 2020 Promulgating National climate change adaptation plan for 2021–2030 period with a vision by 2050 |

Source: Compilation by the authors

transportation and construction activities. It also defines rights, obligations, and responsibilities of manufacturers, importers of transport equipment and vehicles, as well as those of state management agencies. The Law No. 72/2020/QH14 on Environmental Protection dated November 17, 2020, came into force in January, 2022. Article 65 titled "Environmental Protection in Transport" encourages the use of public transport and fuel-efficient vehicles and outlines a roadmap for converting or phasing out fossil fuels-powed vehicles. This article on green transport presents a new addition that was not inclued in the 2014 version of the law.

Mitigation-related strategies have been developed and implemented over recent years. In the Nationally Determined Contribution to the UNFCCC, Vietnam has committed to reducing GHG emissions by 8% using domestic resources, with the potential to increase this reduction to 25% by 2030 contigent upon international support. One of the key strategic frameworks guiding climate action is the National Green Growth Strategy and Action Plans (Decision No. 1658/QD-TTg 2021), which focuses on three main tasks: (1) reducing GHG emissions intensity per unit of GDP, (2) greenifying economic sectors, and (3) greenifying lifestyle and promoting sustainable consumption. These objectives are backed by specific targets and action plans extending through 2050:

Regarding transport sector, the National Action Plan focuses on the following points:

- Formulate and revise policies for the development of green traffic infrastructure facilities and public transport systems, and improvement of the freight transportation productivity in transportation sub-sectors.

- Formulate and organize the implementation of solutions to encourage the use of vehicles powered by clean, economical and efficient energy, as well as eco-friendly technologies; promote the restructuring of transportation market share by shifting cargo transport from road to inland waterway, sea and railway.
- Prioritize use of resources for investment in, finishing and operation of green traffic infrastructure facilities in a manner that ensures economic efficiency and environmental protection, contributes to reduction of greenhouse gas emissions, and improves resilience to climate change and sea level rise. Implement programs on research and application of science and technology to ensure efficient use of energy in investment projects on improvement of public transport infrastructure, development of high-volume public transport infrastructure, non-motorized transport infrastructure and infrastructure facilities that ensure connection between different modes of transport.
- Review and propose amendments to schemes/projects on development of transport sector in a manner that ensures restructuring of transport market share towards green and sustainable growth. Do research on formulation and implementation of national programs/schemes on development of eco-friendly means of transport and eco-friendly public transport systems, which include electric vehicles, development of green logistics centers and green ports, and use of new technologies and new energy to replace traditional fuel for transportation vehicles and equipment.
- Do research and apply digital technology to optimize the management and operation of traffic infrastructure facilities and transport activities, ensuring safe and smooth traffic flows while reducing fuel consumption.

### 3.1.2 Strategies for Sustainable Transport

The Ministry of Transport has been assigned as the national focal agency responsible for issuing and implementing policies, programs, and plans related to the transport sector. In 2013, the Government issued Decision No. 355/QD-TTg, which approved "Adjusted Strategy for Development of Vietnam's Transport through 2020, with an Orientation Towards 2030." This strategy aims to establish a nationwide transport network that leverages the country's geographical advantage and natural resources. In addition, it highlights Vietnam's commitment to international integration through initiatives such as the development of Trans-ASEAN railway system and the expansion of the regional connectivity within the Mekong subregion.

In alignment with decision from government, MOT proposes Action plan for Sustainable Development for 2013–2023 period (Decision. 4088/QĐ-BGTVT 2013), in which its primary tasks are stated as follows:

- Enhance institutional management capacity.
- Mobilize all resources to effectively execute transportation strategies.
- Improve the quality of the transportation system.
- Encourage the efficient utilization of natural resources.

**Table 8** Policies for sustainable transportation for different transport modes

| Transport mode | Related policy | Objectives |
|---|---|---|
| Roadway | Joint circular No. 43/2014/TTLT-BGTVT-BCT 2014 "Regulation on energy labeling for the types of automobiles from 07 seats or less" | Provide transparent information about types of fuel, fuel consumption of the automobile which is inspected, certified, or registered to help consumers identify and choose the fuel-saving automobile |
| | Decision 16/2019/QĐ-TTg 2019 "Exhaust emission standards for new and in-use vehicles" | Provide the roadmap for application of the maximum permitted limits of exhaust gases to off-road vehicles, based on the national standard TCVN 6438:2018. Compared to Decision 249/2005/QĐ-TTg, this updated regulation has tightened the emission standards for in-use motor vehicle |
| Airway | Decision 4206/QĐ-BGTVT 2016 "Promulgating action plans to reduce $CO_2$ emissions in civil aviation in Vietnam for the period of 2016–2020" | Build a sustainable air transport network through aircraft modernization, promotion of low-carbon equipment and facilities, and route optimizations |
| | Circular No. 22/2020/TT-BGTVT "Management of fuel use and $CO_2$ emission of aircrafts in civil aviation operation" | Prescribe collection, monitoring methods and report of fuel use and $CO_2$ emission of aircrafts |
| Railway | Decision No. 214/QD-TTg 2015 "Approving adjustments to the development strategy for Vietnam's railway transport up to 2020 with a vision toward 2050" | Increase modal share of railway to 5–8% for passenger transport and 5–6% for freight transport by 2050 through network and infrastructure development and vehicle modernization |

Source: Compilation by the authors

- Strengthen pollution control measures in transport operations.
- Promote the adoption of technologies aligned with sustainable development.
- Raising public awareness about sustainable transport development (Table 8).

In addition, Vietnamese government authorities, in collaboration with ministries, research institutes, and international organizations hold propaganda sessions on newspapers, conferences, and seminars to raise public awareness about sustainable development and its mutual benefits to the company and the community. An example of success is the "Sustainable Freight Transport and Logistics in the Mekong Region" project, funded by European Union, which has notably gained remarkable achievement in reducing fuel consumption by heavy trucks in freight transport. Over the course of 36 months (first February, 2016–31st January, 2019), 100 SMEs were trained on eco and defensive driving. As a results, fuel consumption was reduced by 11.7% per 100 km for empty heavy-duty trucks and by 11.4% for loaded trucks. Consequently, CO2 emissions dropped by approximately 11.4–11.7% (Switch Asia, 2019).

### 3.2 Response from Companies

DHL Group takes the lead in going-green solutions in the world and also in Vietnam. In 2008, DHL became the first logistics company to set targets for environmental and climate change protection: net zero carbon emissions by 2050 through a well-known program "GoGreen". DHL believes that offering sustainable and eco-friendly services can enhance the company's long-term competitiveness and these environmental protection campaigns act as a green marketing strategy, which could develop a positive brand image. On the one hand, it attracts green customers, thereby increasing sales and revenue; while on the other hand, the project also contributes to sustainable growth—a win-win outcome for business and environment.

Their key sustainable initiatives include optimizing transportation routes, utilizing eco-friendly vehicles, and promoting alternative fuels. DHL Group pursues clear, measurable environmental, social and responsible governance goals as part of its Strategy 2025. In line with the Science Based Targets initiative, logistics-related greenhouse gas emissions (GHG emissions) are to be reduced to less than 29 million tons of CO2 equivalents (CO2e) by 2030. In 2023, DHL Group had around 35,200 e-vehicles (2022: approximately 27,800) and over 25,000 e-bikes, e-trikes and cargo bikes in use for pick-up and delivery. In addition, further multi-year contracts were concluded for Sustainable Aviation Fuel (SAF) (Meyer, 2024). DHL's ultimate goal is to strike a balance between commercial viability, environmental performance, and operational feasibility.

Additionally, some other companies have launched its long-running campaign such as Kuehne & Nagel with "CO2 Offsetting" and Maersk with "Towards a zero-carbon future." Despite setting a positive logistics trend, these campaigns are still limited to world-class companies. Local logistics enterprises are mostly small and medium ones and their application of green supply chains and green practices has been insignificant.

## 4 Lessons Learnt from China for Vietnam: In-Use Road Vehicle Emission Control Programs

Over the past three decades, China has seen a significant rise in vehicle ownership, driven by rapid socioeconomic development and urbanization. The country's GDP per capita in 2023 has increased by 38 times that of 1990 and about 66.2% of the total population residing in urban areas (World Bank, 2020). Over the same period, the registered vehicle population surged from 5.5 million in 1990 to 329.2 million units in 2023. This rapid motorization has presented significant challenges, including energy security concerns, worsening urban and regional air pollution, and contributions to global climate change. According to the International Energy Agency (IEA), China's roadway sector released 565 million tons of CO2 equivalent to 4% of global fossil fuel CO2 emissions (IEA 2024). China has initiated its vehicle emission control programs for road vehicles since the 1990s, two decades later than other highly motorized regions such as the United States and Europe, but gained significant successes.

## 4.1 *Ensuring Emission Standards Compliance of On-Road Vehicles: Inspection and Maintenance*

Inspection and maintenance programs mandate periodic safety and emissions testing for vehicles, with the frequency depending on the vehicle's type and age—typically once or twice a year. During these inspections, technicians evaluate not only the vehicle's emissions but also its compliance with type-approval standards for key components. Vehicles that fail the tests must be repaired or maintained to meet the required standards before being permitted to operate on the road again (Table 9).

The policy is regarded as a preparation for other emissions reduction methods listed in below sections.

## 4.2 *Reducing Emissions Through Vehicle Management*

### 4.2.1 Vehicle Environmental Labelling Policy

After national emission standards came into force, China introduced an environmental labeling policy to identify high-emission vehicles within the in-use fleet. Local environmental protection bureaus issue these labels, which are distributed to vehicle owners after periodic inspection tests. Typically, pre-China I light duty vehicles and pre-China III heavy duty vehicles are given yellow labels. In Beijing, vehicles meeting China V emission standards are awarded blue labels, while green labels are issued to vehicles that fall between the yellow and blue categories. The

**Table 9** Inspection/maintenance for some vehicle categories

| | Vehicle age | | | | | | | | | | | | | | | | | | | |
|---|---|---|---|---|---|---|---|---|---|---|---|---|---|---|---|---|---|---|---|---|
| | 1 | 2 | 3 | 4 | 5 | 6 | 7 | 8 | 9 | 10 | 11 | 12 | 13 | 14 | 15 | 16 | 17 | 18 | 19 | 20 |
| Light-duty passenger vehicle | Exempt | | | | | | Every year | | | | | | | | | Twice a year | | | | |
| Commercial passenger car | Every year | | | | | | Twice a year | | | | | | | | | | | | | |
| Commercial truck | Every year | | | | | | | | | | Twice a year | | | | | | | | | |
| Medium/heavy-duty passenger vehicle | Every year | | | | | | | | | | Twice a year | | | | | | | | | |
| School bus | Twice a year | | | | | | | | | | | | | | | | | | | |
| Others | Every year | | | | | | | | | | | | | | | | | | | |

Source: Compilation by the authors

**Table 10** Label colors for vehicles in Beijing, China

| | | | | | | | |
|---|---|---|---|---|---|---|---|
| | | | | | | | |
| Light Duty Vehicles | | | | | | | |
| | | | | | | | |
| Heavy Duty Vehicles | | | | | | | |
| | | | | | | | |
| | | Uncontrolled | China I | China II | China III | China IV | China V |

Source: Compilation by the authors

environmental labeling policy is considered a "low-hanging fruit" policy which helps environmental protection authorities distinguish different vehicle groups. It also facilitates targeted measures for high emitting vehicles, such as traffic restrictions and subsidized scrappage programs, to address emissions effectively (Table 10).

### 4.2.2 Scrappage and Replacement

The scrappage program has been implemented since the 1990s, targeting yellow-labeled vehicles in Beijing. Vehicle owners who participate in the program are eligible to receive incentives for scrapping their yellow-labeled vehicles early or transferring them outside of Beijing. To be more specific, between September 27, 2008 and June 30, 2009, participants received compensation ranging from 800 RMB to 25,000 RMB (approximately \$131 to \$4,086). For the participation on second half of the year 2009, lower incentives were provided, but the program was extended through 2010 with slightly increased subsidies and the subsidy programme is valid until 10 January 2025.

During 2 years (2009–2010), the program has successfully phased out 150,000 yellow-labeled vehicles. This effort led to significant environmental benefits, including daily reductions of nearly 32 tons of $NO_x$ emissions, 245 tons of CO emissions, 35 tons of HC emissions, and 3 tons of PM emissions (World Bank, 2019).

### 4.3 Reducing Emission Through Traffic Restriction

In addition to implementing technology-driven emission control programs and regulations, traffic restrictions serve as a supplementary measure to lower emissions from in-use vehicles. Traffic congestion, which reduces vehicle speeds in urban areas, contributes to increased carbon emissions. In Beijing, yellow-labeled vehicles had been prohibited within the Second Ring Road (around 60 $km^2$) in 2003. The restriction area was then extended to within the Sixth Ring Road (around 2500 $km^2$) in 2009 and eventually covered the entire municipal boundary of Beijing (about 16,000 $km^2$) in 2015. While the program had minimal impact on overall traffic demand, it partly accelerated the replacement of older, more polluting vehicles, thereby reducing emissions.

Even though these programs have successfully cut down transport-related carbon emissions, there are several challenges that make China's transport sector still a major emitter in the country. Firstly, the rapid growth in the number of vehicles has outpaced the development of transportation infrastructure. As a result, the road network lacks the capacity to support the vehicle industry boom, leading to traffic congestion and avoidable greenhouse gas (GHG) emissions. Secondly, there is a disconnect between the upgrading of fuel standards and vehicle emissions standards. Specifically, fuel standards have not kept pace with vehicle emissions standards. The China III gasoline standard, implemented in January 2010, came two and a half years after the China III light-duty vehicle standard. Likewise, the China III diesel standard was rolled out nationwide in July 2011, four and a half years after the China III heavy-duty diesel vehicle standard was enforced. This lag in fuel standard upgrades undermines the effectiveness of emissions standards, limiting their impact on reducing pollution.

Vietnam is supposed to follow the same socio-economic growth rate as China. Firstly, both countries experienced growth accelerations starting from comparable per-capita income levels. Secondly, by 2013, Vietnam had caught up with China, reaching a similar point in its growth acceleration after 23 years. However, Vietnam's energy system has carbonized even faster than China's (Zimmer et al., 2015). Acknowledging that Vietnam also meets the same obstacles as China in their process of cutting GHG emissions, these aforementioned case studies could act as references for the country to issue its transport-related emission controls strategies.

## 5 Recommendations for Vietnam to Improve Green Transportation Performance

The Government of Vietnam has been working towards sustainable transportation through various initiatives. After comprehensively assessing operations of Vietnam's transport sector, the study proposes some GHG mitigation strategies which align with the vision of the Ministry of Transport (Table 11).

**Table 11** GHG Mitigation Strategies in the Transport Sector

| | |
|---|---|
| Energy Efficiency Improvement | Fuel economy standards |
| | Fleet modernization |
| Fuel shift | Promotion of biofuels |
| | Promotion of electricity |
| Modal shift | Passenger modal shift from private vehicles to public transportation |
| | Freight modal shift from road to inland waterway transport and coastal shipping; railway |

Source: Compilation by the authors

## 5.1 *Improve energy efficiency*

### 5.1.1 Build Fuel Economy Standards

Fuel economy currently plays a limited role in vehicle purchasing decisions in Vietnam, with four mandatory fuel consumption standards for automobiles and two voluntary standards for two-wheeled mopeds and motorcycles. To promote manufacturing of fuel-efficient vehicles, it is essential to conduct a comprehensive analysis of the vehicle market, establish reliable testing procedures, and engage stakeholders to ensure that fuel economy standards are equitable and result in tangible cost savings for consumers.

Vietnam is aligning with the Chinese fuel economy standard, albeit with a five-year lag. This approach minimizes costs for automakers who already have placed their supply chains. However, these standards should be regularly reviewed and strengthened to drive innovation among manufacturers and encourage the development of more efficient solutions. This approach not only supports environmental goals but also fosters healthy competition among manufacturers, enhancing market competitiveness (Table 12).

### 5.1.2 Fleet Modernization

To mitigate the GHG emissions, a fleet modernization program is proposed to encourage vehicle owners to replace outdated vehicles with newer, environmentally friendly alternatives.

*For roadway*, the program could be specified into two initiatives:

- Policy to restrain the use of old trucks through additional charges

This policy proposes implementing a surcharge on road usage fees based on the age and emissions of vehicles. Older trucks with higher emissions would be subject to higher surcharges, reflecting their disproportionate environmental impact. This measure could be integrated into the existing Circular No 293/2016/TT-BTC, regulating the road usage charges by introducing differentiated fee structures in accordance with vehicle age.

**Table 12** TCVN 6438:2018 - Emission limits of off-road vehicles

| Pollutants in exhaust gas | Vehicles fitted with spark-ignition (gasoline) engines | | | | | | Vehicles fitted with compression ignition (diesel) engines | | | |
|---|---|---|---|---|---|---|---|---|---|---|
| | Automobiles | | | | Mopeds, Motorcycles | | | | | |
| | Limit 1 | Limit 2 | Limit 3 | Limit 4 | Limit 1 | Limit 2 | Limit 1 | Limit 2 | Limit 3 | Limit 4 |
| CO (%volume) | 4.5 | 3.5 | 3.0 | 0.5 0.3 | 4.5 | | – | – | – | |
| HC (ppm volume) | | | | | | | | | | |
| - Four-stroke engine | 1200 | 800 | 600 | 300,200 | 1500 | 1200 | | | | |
| - Two-stroke engine | 7800 | 7800 | 7800 | 7800 | 10,000 | 7800 | | | | |
| - Special engine | 3300 | 3300 | 3300 | 3300 | | | | | | |
| Lambda (λ) | | | | 0.97–1.03 | | | | | | |
| Smoke opacity (%HSU) | | | | | | | 72 | 60 | 50 | 45 |

Source: Compilation by the authors

- Policy to encourage buying green vehicles

The proposed policy suggests that the government should give monetary benefits for scrappage of outdated vehicles and incentivize the purchase of environmentally friendly alternatives that meet the current Euro 4 emissions standard.

Discounts for scrapping older vehicles: customers who scrap their trucks at authorized recycling centers would receive a portion of the vehicle's value as a discount when purchasing a new truck.

Incentives for purchasing new vehicles: customers who purchase green vehicles would get exemption for a portion of the registration tax.

*For the inland waterway and maritime transport sectors,* Europe follows a general trend of achieving "higher efficiency with fewer vessels." In contrast, Vietnam's national fleet is predominantly composed of small vessels, with typical cargo vessel sizes ranging from 100 to 250 DWT. These are significantly smaller than those in European countries with well-established inland waterway transport (IWT) networks, where vessels average around 2500 DWT. Additionally, the Vietnamese fleet is aging, with an average vessel age exceeding 30 years. This results in a large number of vessels operating with lower efficiency. To address these challenges, a policy proposal recommends the establishment of a maritime development fund to support the financing, construction, and acquisition of new, larger vessels while discouraging the use of smaller vessels. The policy also emphasizes the importance of sustainable shipbuilding practices, promoting clean engine technologies, and encouraging fuel-efficient operations. Such initiatives are crucial, as greenhouse gas (GHG) emission reductions from fleet modernization are expected to surpass those achieved by shifting transportation from roads to waterways.

## 5.2 *Fuel Shift*

### 5.2.1 Promotion of Biofuels

Currently, the most commonly used biofuel in Vietnam is E5, which consists of 95% gasoline and 5% ethanol. However, the current environmental protection tax is caculated based on the gasoline content. Specifically, E5 is taxed at 95% of the standard gasoline tax rate (i.e., 1,900 VND/liter), while pure gasoline is taxed at 2,000 VND/liter. Although E5 offers significant environmental benefits, the tax exemption amount is not attractive enough. Thus, reducing the environmental tax on gasoline (including ethanol) should be considered to create a greater difference in tax reduction, incentivizing vehicle owners to use green energy.

### 5.2.2 Promotion of Electricity

To achieve zero carbon emissions, promoting electric vehicles, such as electric motorbikes and cars, is essential in the long term. Currently, electric motorbikes on the market have an affordable initial purchase price, ranging from VND 7 million to VND 15 million, depending on the model. However, electric motorbikes can end up being more expensive than traditional gasoline-powered motorcycles due to their shorter lifespan and higher maintenance costs. Similarly, electric cars still experience modest sales, even in developed countries. Supporting the growth of electric vehicles requires investment in charging infrastructure. While most electric motorbikes will be charged at home, around 50% of electric cars will depend on public charging stations. International experience shows the need to develop one fast-charging station for every 100 electric cars, with each station costing between US$10,000 and US$20,000. Although this initiative requires upfront investment in both electric vehicles and infrastructure, it ultimately results in reduced fuel consumption and long-term savings on operational and maintenance costs.

## *5.3 Modal Shift*

### 5.3.1 Passenger Modal Shift from Private Vehicles to Public Transportation

Promoting the development of buses and metro systems takes a crucial role in reducing greenhouse gas emissions in the transportation sector. This shift involves encouraging passengers to move from motorcycles and private cars to public transit, which consumes significantly less energy per passenger-kilometer. However, the current public transportation infrastructure in metropolitan cities like Hanoi and Ho Chi Minh City has not yet met the growing travel demand. The progress of construction and operation for mass transport (BRT, metro rail transit) missed the deadline. For example, three MRT lines are currently under construction and inspection. In Hanoi, Line 2A, running from Cat Linh to Ha Dong, planned to operate in 2019, but the actual first trial was in on 2020. Line 3 (Nhon - Hanoi) and Line 1 (Ben Thanh - Suoi Tien, Ho Chi Minh City) are projected to start operations in 2022. The pending in infrastructure development will result in procrastination of many subsequent plans. Therefore, the Government of Vietnam should prioritize financial and technical assistance to complete these master projects in time. A balanced "push and pull" strategy is essential, combining incentives for public transit use with policies that restrict private vehicle usage, such as: extra charges for private automobiles entering central areas, environmental protection tax on outdated vehicles, building dedicated lanes for BRT, … While the primary driver for mass transit adoption is its potential to save travel time and enhance comfort, it also contributes to reducing greenhouse gas emissions, aligning with environmental goals of Vietnam.

**Table 13** Proposed Interventions for developing port connectivity in Red Delta Region and its projected impact on CO2 emissions

| Intervention name | Description | Change in IWT/ coastal volume (tons per day) | CO2 emissions reduction (tons per day) |
|---|---|---|---|
| Upgrade waterway Corridor 1 of the Red River Delta | Raises Corridor 1 (Quang Ninh–Haiphong– Pha Lai–Hanoi–Viet Tri) from waterway Class II to Class I | 3623 | 414 |
| Upgrade waterway Corridor 2 of the Red River Delta | Raises Corridor 2 (Haiphong–Ninh Binh) from waterway Class III to Class I | 1497 | 202 |
| Upgrade waterway Corridor 3 of the Red River Delta | Raises Corridor 3 (Hanoi–Day/ Lach Giang) from waterway Class III to Class I | 543 | 55 |

Source: World Bank (2013)

### 5.3.2 Freight Modal Shift from Road to Inland Waterway Transport and Coastal Shipping; Railway Transport

Given that roadway transport is the most carbon-intensive mode of transportation, shifting freight movement from roads to waterways and railways is essential for transitioning toward a low-carbon economy.

In order to quickly promote waterway transport, investments in seaports, river ports and transport linkages are required nationwide, including:

- Reconfiguring existing terminals, deepening channels and berths.
- Improving inland access by road, rail and barge.
- Speeding up vessel loading and unloading time, since air quality would be enhanced by reduction of vessel dwelling time.
- Developing connectivity between seaports and IWT ports in both Red Delta ang Mekong Delta (Table 13).

- Transitioning to rail transport will demand significant investments in railway infrastructure and train carriages. This involves developing and upgrading railway freight terminals and ICDs, expanding railways and accessible ports, and improving cargo handling facilities. Key areas for improvement are detailed below:
- For existing railway network:

Step-by-step improving the Hanoi – Ho Chi Minh City railway, spanning around 1726 km to achieve faster average speeds. This includes the reinforcement and replacement of weak bridge and basement, the transition to double-track railway of 1435 millimeters size, the addition of railway lines within stations, and the extension of station tracks.

It is necessary to renovate and upgrade railway stations along key transport corridors to enhance service quality, with a particular focus on the Hanoi - Ho Chi

Minh City route, as well as Hanoi (Gia Lam) – Hai Phong (96 km), Hanoi (Yen Vien) – Lang Son (156 km), and the Hanoi (Yen Vien) – Lao Cai (285 km lines).

- For newly constructed railway:

The construction of the Yen Vien - Lac Dao - Ngoc Hoi route, considered as east belt line of, approximately 80 km) within Hanoi's hub areas, is planned alongside the railway connection between Hai Phong seaport and Lach Huyen (32.65 km), and Bien Hoa - Vung Tau railway route, with priority given to the Trang Bom – Thi Vai section, 65.4 km long.

Additionally, the construction of railway lines such as Hanoi – Lang Son, Lao Cai – Ha Noi - Hai Phong, Ho Chi Minh City - Can Tho (with priority on the Ho Chi Minh City - My Tho section). Ho Chi Minh City - Can Tho, Di An - Loc Ninh and Dak Nong – Chon Thanh is planned to enhance connectivity with the Trans-Asia railway network, along with the 67kmVung Ang - Cha Lo (Mu Gia) line which willconnect Vietnam's railway system with Laos.

## References


Asian Development Bank (ADB) (2013) Vietnam: environment and climate change. Asian Development Bank. Available online at: https://www.adb.org/sites/default/files/institutional-document/33916/files/viet-nam-environment-climate-change.pdf

Chinowsky, P. S., Schweikert, A. E., Strzepek, N., & Strzepek, K (2015) Road Infrastructure and Climate Change in Vietnam. Sustainability, 7(5), 5452–5470. https://doi.org/10.3390/su7055452

Decision No 804/2022/QĐ-TTg dated July 8, 2022 was issued by the Prime Minister promulgating the list of Vietnam's sea ports.

Department of Climate Change (2021) The impact of climate change on Vietnam's agricultural economy, Available online at: https://pilot.dcc.gov.vn/en/climate-change-adaptation/the-impact-of-climate-change-on-vietnam-1535#:~:text=It%20is%20forecasted%20that%20by,in%20central%20and%20southern%20regions

Espagne E. (ed.), T. Ngo-Duc, M-H. Nguyen, E. Pannier, M-N. Woillez, A. Drogoul, T. P. L. Huynh, T. T. Le, T. T. H. Nguyen, T. T. Nguyen, T. A. Nguyen, F. Thomas, C. Q. Truong, Q. T. Vo, C. T. Vu. 2021. Climate change in Viet Nam; Impacts and adaptation. A COP26 assessment report of the GEMMES Viet Nam project. Paris. Agence Française de Développement. Available online at: https://horizon.documentation.ird.fr/exl-doc/pleins_textes/2022-05/010084119.pdf

International Energy Agency (2024), Available online at: https://www.iea.org/energy-system/transport.

Lam Y Y, Sriram K, Khera N (2019) Strengthening Vietnam's trucking sector towards lower logistics costs and Greenhouse Gas Emissions. World Bank. Available at https://documents.worldbank.org/en/publication/documents-reports/documentdetail/090224b086b75033_2_0 (Accessed: 01 October 2020)

Meyer, T (2024) Resilient business model: DHL Group achieves annual targets for 2023 in a weak global economic environment. Available at https://group.dhl.com/en/media-relations/press-releases/2024/dhl-group-annual-earnings-2023.html

Ministry of Industry and Trade (2019) Vietnam Logistics Report 2019. Available at: https://gosmartlog.com/wp-content/uploads/2019/12/Bao-cao-logistics-viet-nam-2019.pdf

Nguyen, K (2024) Transport segments face up to GHG requirements. Vietnam investment review. Available online at: https://vir.com.vn/transport-segments-face-up-to-ghg-requirements-114349.html

Nguyen MQ, Pham TTX, Phan TTH (2019) Traffic Congestion - A Prominent Problem in Vietnam Current Situation and Solutions. European Journal of Engineering Research and Science, 4(9): 112–116, https://doi.org/10.24018/ejers.2019.4.9.1524

Planète Energies (2023, December 27) Global transportation sector CO2 emissions rise. Planète Energies. Available online at: https://www.planete-energies.com/en/media/article/global-transportation-sector-co2-emissions-rise

SLOCAT (2023) Transport, Climate and Sustainability Global Status Report – 3rd edition. www.tcc-gsr.com. Available online at: https://slocat.net/achieving-a-green-energy-transition-in-vietnams-transport-sector-through-electric-mobility

Switch Asia (2019) Sustainable freight and logistics in the Mekong region: Impact sheet. Switch Asia. Available online at: https://www.switch-asia.eu/site/assets/files/1518/sustainable_freight_and_logistics_mekong_impact_sheet.pdf

Tiseo I (2024) Global greenhouse gas emissions 1970-2023. Statistica. Available online at: https://www.statista.com/statistics/1285502/annual-global-greenhouse-gas-emissions/#:~:text=In%202023%2C%20global%20greenhouse%20gas,carbon%20dioxide%20equivalent%20(GtCO%E2%82%82e)

VCCI (2024) Vietnam, China cross-border railway project to start construction in 2025. Available at: https://en.vcci.com.vn/vietnam-china-cross-border-railway-project-to-start-construction-in-2025 Vietnam, China cross-border railway project to start construction in 2025

Vietnamnet (2024) Non-communicable diseases account for 80% of deaths in Vietnam. Available online at: https://vietnamnet.vn/en/non-communicable-diseases-account-for-80-of-deaths-invietnam-2349818.html

Wang Y, Zhao S, Wei Y, Li K, Jiang X, Li C, Ren C, Yin S, Ho J, Ran J, Han L, Zee B C-y, Chong KC (2023) Impact of climate change on dengue fever epidemics in South and Southeast Asian settings: A modelling study. Infectious Disease Modelling 8(3):645–655

World Bank (2010) The costs to developing countries of adapting to climate change – new methods and estimates: the global report of the economics of adaptation to climate change study. Consultation Draft, Washington, DC. Available online at: https://documents.worldbank.org/en/publication/documents-reports/documentdetail/667701468177537886/the-costs-to-developing-countries-of-adapting-to-climate-change-new-methods-and-estimates-the-global-report-of-the-economics-of-adaptation-to-climate-change-study

World Bank (2013) China 2030: building a modern, harmonious, and creative society. World Bank. Available online at: https://documents1.worldbank.org/curated/en/781101468239669951/pdf/China-2030-building-a-modern-harmonious-and-creative-society.pdf

World Bank (2019) The changing Nature of Work. Available online at: https://documents1.worldbank.org/curated/en/195101548767451651/pdf/Overview.pdf

World Bank (2020, February 6) Vietnam's long-term growth performance in comparative perspective. World Bank Blogs. Available online at: https://blogs.worldbank.org/en/eastasiapacific/vietnam-s-long-term-growth-performance-comparative-perspective

World Bank (2021) Climate risk country profile Vietnam. Available at: https://climateknowledgeportal.worldbank.org/sites/default/files/2021-04/15077-Vietnam%20Country%20Profile-WEB.pdf

World Bank (2022) Vietnam fiscal impact study. World Bank. Available online at: https://documents1.worldbank.org/curated/en/490331468190155446/pdf/97472-BRI-Box391476B-PUBLIC-Vietnam-Fiscal-Impact-Study-Final.pdf

World Bank (2023) GDP by countries 2023. Available online https://data.worldbank.org/indicator/NY.GDP.MKTP.CD

Zimmer, A., Jacob, M., Steckel, J. C (2015) What motivates Vietnam to strive for a low-carbon economy? — On the drivers of climate policy in a developing country. Energy for Sustainable Development, 24(3):19–32. Available at: https://doi.org/10.1016/j.esd.2014.10.003

**Assoc. Prof. Tuan-Anh Bui** (PhD in Economics, National Economics University, Vietnam) is a lecturer in charge of Human Resource Management at Foreign Trade University. His research focuses on organizational behavior, innovation, inclusive growth, industrial pollution, business linkage, and woman entrepreneurship.

**Huong-Thu Le** is currently working as a business analyst for FPT Software Company Limited. She holds a Bsc in Economics from the Foreign Trade University, Hanoi, Vietnam.

**Hanh-Thi Vu** (Ph.D in Economics and Management, Université Libre de Bruxelles, Belgium) is a lecturer in International Business, Foreign Trade University, Vietnam. Her research interests include sustainable trade, international business, and waste management.

# Part II
# Economic Sustainability

# The Impact of Foreign Direct Investment on Environmental Quality

**Vinh-Thi Hong Cao and Dung-Tien Tran**

**Abstract** In accordance with the increasing role of international capital flows, especially foreign direct investment (FDI) for host countries, a rising concern about if this kind of capital could actually help countries to achieve their sustainable development aims. Besides undoubted benefits, especially for developing countries (such as bridging the gap of capital demand and supply for economic growth) that FDI could bring, there have been many cases of FDI that make the environment quality worse. This will for sure significantly affect the possibility of host countries to gain sustainable development goals. For that reason, this chapter attempts to estimate the impact of FDI on environmental quality (EQ) across a broad dataset of 230 countries worldwide. Applying the panel data techniques of fixed and random effects, we have found out the negative impact of FDI on EQ, proving the hypothesis of "pollution haven." In addition, other results are also discovered, in particular: (i) based on the environmental Kuznets curve theory, there is an inverted U-shaped relationship between pollution levels and national income and (ii) the greater openness to trade leads to worse EQ, as the emissions of both CO2 and GHG increase.

## 1 Introduction

FDI inflows have significantly contributed to the development process of host countries. However, despite benefits such as capital acquisition and budget deficit reduction, FDI causes negative effects on the local environment, one of the three key pillars of sustainable development. As countries increasingly care more about the

V.-T. H. Cao (✉)
Foreign Trade University, Dong Da, Ha Noi, Vietnam
e-mail: caovinhftu@ftu.edu.vn

D.-T. Tran
Urban Development Agency, Ministry of Construction, Hai Ba Trung, Ha Noi, Vietnam
e-mail: dungtt@moc.gov.vn


H.-T. Vu, L.-P. Dana (eds.), *Vietnam on the Road to Sustainable Economic, Social and Business Development*, Frontiers in South and Southeast Asian Development Research, https://doi.org/10.1007/978-981-96-4371-4_5

aim of sustainable development, they would look more closely into the real effects of FDI on different perspectives of host countries. One of the starting things taken into remarkable consideration is environmental quality.

*What is the actual effect of FDI on the EQ of the countries who receive FDI, the investors of which are private ones with the final aim for their investment activities to achieve as high profit as possible?* On one hand, investment activities that incorporate modern, environmentally friendly technology and produce clean products for the host country's markets can help reduce pollution, leading to positive environmental outcomes. On the other hand, because FDI inflows are primarily profit-driven and investors often overlook issues such as wastewater, toxic gases, and industrial waste disposal, they can pose significant risks of environmental degradation. As a result, the answer to the above question is still ambiguous.

For that reason, what we would like to clarify is the impact of FDI on EQ of a wide range of 230 countries for a period of 1970–2019 using the panel data techniques of fixed and random effect. In addition, as looking into the effect on EQ, we also want to check if there exists the influence of national income on that variable as mentioned in the environmental Kuznets curve theory.[1]

The chapter are structured as follows: Section 2 refers to the Literature review, then Sects. 3 and 4 present an overview of FDI and EQ, and theoretical impacts of FDI on EQ. Then, Sect. 5 focuses on the quantitative analysis of the effects of FDI on EQ. The material presented in Sect. 6 constitutes the final conclusion.

## 2 Literature Review

There has been a considerable amount of research on the impact of FDI inflows on EQ, with most studies highlighting negative effects. For instance, Lee (2009), using Granger causality methods, found that FDI had an impact on CO2 emissions in Malaysia during the period from 1970 to 2000. Similarly, Pao and Tsai (2011) examined the effect of FDI on CO2 emissions using panel data from Russia, Brazil, India, and China between 1980 and 2007.

Dinh and Lin (2014), in their study on the relationship between CO2 emissions, energy consumption, economic growth, and FDI in Vietnam using data from 1980 to 2010 and various analytical methods, found that lax environmental regulations helped attract more FDI to Vietnam, which in turn contributed to higher levels of environmental pollution.

In their paper titled "FDI, Growth and the Environment: Impact on Quality of Life in Malaysia" published in Procedia—Social and Behavioral Sciences, Hitam and Borhan (2012) explored the relationship between GDP growth and environmental degradation in Malaysia from 1965 to 2010. Using a nonlinear model, the authors examined how foreign direct investment (FDI) affected environmental degradation

[1] See Appendix 1 for the clarification of Environmental Kuznets Curve theory

during this period. The findings revealed the presence of the environmental Kuznets curve, with FDI contributing to increased environmental degradation in Malaysia.

To et al., (2019) investigated the relationship between foreign direct investment (FDI) and environmental degradation, while also testing the validity of the traditional environmental Kuznets curve (EKC) in the context of emerging Asian markets. Using data from these countries spanning 1980 to 2016, the study confirms the existence of the EKC in the region and demonstrates that FDI has a significant impact on the environment.

However, some studies suggest that FDI does not have a negative impact on EQ or the Kuznets curve. A notable example is the research by Berna Kirkulak et al., (2011), who analyzed the effect of FDI on air pollution in China, using data from 286 cities between 2001 and 2007. The study aimed to better understand the impact of economic growth and foreign investment on environmental sustainability. The authors found that FDI did not negatively affect air quality in China, rather, the presence of FDI was associated with a reduction in air pollution. This outcome can be attributed to the role of FDI as a key source of advanced technology in China.

# 3 Overview of FDI and Environment Quality

## *3.1 Foreign Direct Investment*

### 3.1.1 Definitions and Characteristics

The definitions of FDI have been addressed by various organizations and incorporated into the laws of different countries. Among these, one of the most widely accepted definitions comes from the International Monetary Fund (IMF) (1993) in the Balance of Payments Manual, fifth edition, paragraph 359. According to the IMF (1993), FDI is defined as "the category of international investment that reflects the objective of a resident entity in one economy obtaining a lasting interest in an enterprise resident in another economy." The IMF further explains that the goal of establishing a "lasting interest" refers to the intention to "ensure a significant degree of influence by the direct investor." To achieve this long-term relationship, the IMF specifies that the investor must own 10% or more of the ordinary shares or voting power (in the case of an incorporated enterprise) or the equivalent for an unincorporated enterprise. This definition aligns with the OECD's Detailed Benchmark Definition of Foreign Direct Investment, as stated in the Balance of Payments Manual, fifth edition.

Based on the definition provided above, several key characteristics of FDI can be identified (Vu Chi Loc, 2012). First, the primary goal of investment through FDI is financial gain (such as profit), rather than social benefits. While some FDI investors may engage in corporate social responsibility (CSR) activities, their main focus remains on achieving financial returns, as they are private investors. As private investors, FDI investors typically have autonomy over the nature of their investment activities, including the amount of capital, sector of investment, business strategy, and human resource allocation.

The second characteristic highlighted by the definition is the long-term interest of FDI investors in the enterprises they invest in, indicating a desire to establish a sustainable relationship that requires significant influence in management. Another important point is that FDI investors hold controlling rights, typically based on voting power, which are essential for exerting influence. As per the IMF and OECD, investors acquire this control by owning 10% or more of the ordinary shares or voting power (for incorporated enterprises) or the equivalent in unincorporated enterprises.

Finally, FDI contributes to technology transfer, as investors bring advanced technology from their home countries to the host countries. This includes not only physical technology (such as machinery and equipment) but also the sharing of knowledge and the application of modern business practices in the host country's operations.

**Host country determinants of FDI**

Various approaches have been proposed regarding the factors in host countries that can influence FDI investors. In the World Investment Report (UNCTAD, 1998), these determinants were analyzed and categorized into three groups (see Fig. 1), which have been widely referenced by researchers. The categories include (i) policy framework, (ii) economic factors, and (iii) business facilitation.

Policy Framework Factors

These policy framework factors are classified into the inner (core) and outer-ring policies.

Core FDI policies (referred to as the inner ring) encompass various aspects, including rules and regulations governing the entry and operations of foreign investors, the standards of treatment they receive, and the functioning of the markets in which they operate. These policies are seen as directly influencing FDI investors. Specifically, the requirements related to entry and operations significantly impact the ability of FDI investors to engage in investment activities. In some countries, foreign investors are restricted from participating in certain sectors, such as

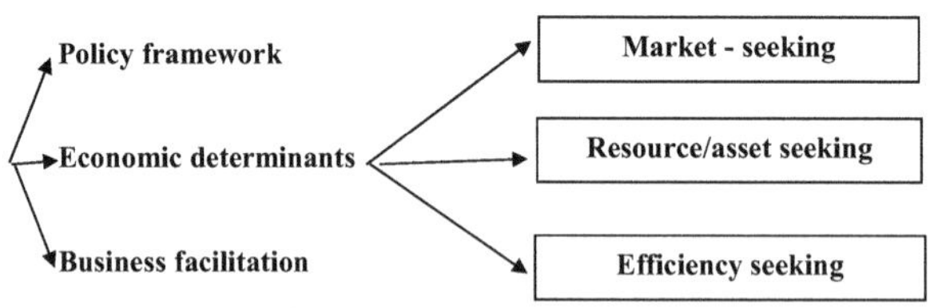


**Fig. 1** Host country determinants of FDI. Source: Summarized by the authors from UNCTAD (1998)

electricity and telecommunications, even though these sectors might be more efficiently managed by foreign investors rather than domestic ones. In addition to the entry, regulations about the operation of foreign investors could create a strict barrier on their willingness to join the market. For example, the provision which just allows FDI investors to establish joint venture, not 100% foreign invested enterprises could make it harder for them to carry out their business independently.

About the *standard of treatment*, there could be two things regarding non-discrimination treatment that are also carefully considered by FDI investors. The first thing regards most-favored nation (MFN) treatment, which focuses on the non-discrimination among foreign investors with different nationalities. Meanwhile, the second thing is about national treatment, which indicates the nondiscrimination between foreign and domestic investors. Both of these are also of interest to FDI investors.

Relating to *the structure of the markets* which investors are operating in, this affects the possibility of investors to compete with other competitors in the markets. Whether the market is under the kind of monopoly, oligopoly, monopolistic competition or perfect competition need to be taken into careful consideration by any new entrants, especially the ones from the outside world. Without such care, it is highly possible that they could not compete with the existing competitors and have to exit soon.

Beside these abovementioned inner ring policies, from Fig. 1, the outer-ring policies could include a variety of policies, which also have an impact on the investment decisions of FDI investors. These policies could range from the ones to maintain economic, political, and social stability to trade, tax, or privatization ones. All of these also play a significant role as their adjustment could lead to a significant change to the result of their business activities.

Moreover, in addition to the domestic policies (which are composed of inner and outer ring ones), the increasing international investment agreements need to be taken much care of as well. As such agreements provide investors with preferential conditions and also protect them from bad treatment, their existence will draw much attention from investors.

### 3.1.2 Economic Factors

According to UNCTAD (1998), economic factors are classified based on the motivations of investors, which include (i) market-seeking, (ii) resource/asset-seeking, and (iii) efficiency-seeking. Among these, market-seeking is the most common motivation for FDI investors, whose primary goal is to serve local and regional markets. With such aim, the economic factors of the host countries that are of investors' interest are market size, income per capita, market growth, possibility of host countries to access regional and global markets, etc. For investors with a resource or asset-seeking motive, their primary focus is on the availability of raw materials, low-cost unskilled labor, skilled labor, and other types of assets (as shown in Fig. 1).

The last type of *efficiency-seeking motive* has been getting more and more popular in host developing countries. For the investors with this kind of aim, the economic factors that draw their most attention are highly related to costs of resources and assets, especially input costs.

### 3.1.3 Business Facilitation Factors

In addition to the policy framework and economic factors mentioned above, business facilitation factors also play a significant role in attracting FDI investors. These typically include promotional efforts, incentives for foreign investors, reducing the "hassle costs" of doing business in host countries (such as minimizing corruption and improving administrative efficiency), and offering amenities that enhance the quality of life for expatriate personnel.

## *3.2 Roles of FDI to Host Countries*

In theory, the impact of FDI on host countries is viewed from two main perspectives: capital widening and capital deepening. As a form of capital flow, FDI can help host countries accumulate more capital, a direct effect known as capital widening. Beyond this, FDI also brings benefits such as technology transfer, improved labor skills, and expanded connections between domestic firms and global networks, all of which can lead to spillover effects that enhance productivity in the host country (this is referred to as capital deepening). The key roles of FDI in terms of capital deepening can be outlined as follows:

**FDI raises income of labor**
FDI can contribute to increasing labor income primarily by creating jobs and enhancing local skills. Investment from foreign countries helps promote businesses in developing nations, generating employment opportunities for local people. Karlsson (2007) noted that in China, FDI positively impacted employment growth, which in turn boosted household income and strengthened the spending power of local residents. Specifically, via FDI, multinational companies bring advanced technologies and managerial expertise to less developed countries. According to Kurtishi-Kastrati (2013), foreign firms provide high-quality training to their employees, and some of these skills are transferred to domestic firms later as workers move to those. As a result, host countries could obtain the "managerial superiority" of multinational companies by learning and imitating their practices. At the same time, employees enjoy higher incomes as they upgrade their skills and pursue better job opportunities.

Income levels also rise as FDI inflows stimulate economic growth, which in turn boosts income levels. Tran Trong Hung (2005) argued that FDI, through its effects on economic growth and employment rates, can help reduce poverty and improve the quality of life for local populations. The rise in economic growth and employment demand is closely linked to the increase in household disposable income, leading to a reduction in the number of people living below the poverty line.

**FDI improves health situation of host countries' citizen**

FDI can influence the health of people in a country through several channels:

The first channel is the increased self-awareness of individuals regarding health issues as their income rises. As people earn more money and become more conscious of the importance of health, they are likely to allocate a larger portion of their disposable income to healthcare services. Additionally, they can use this extra income to purchase higher-quality goods, such as organic food. Greater spending on healthcare and quality goods improves overall well-being, which can gradually lead to higher life expectancy.

FDI can also improve health conditions in host countries by offering higher wages compared to domestic firms and by providing safer working environments and better social services. A safe workplace is a mandatory requirement in developed countries and is expected to be consistently maintained by companies. When foreign firms establish operations in the host country, they tend to place more emphasis on the working conditions of their employees than domestic companies.

**FDI contributes to the improvement of education**

Education is a crucial component of a country's development, and FDI plays a significant role in shaping it. Increasingly, foreign investors view education as a promising sector to invest in. They typically target countries with high demand for global-standard education or those with a desire to send people abroad for studies. By investing in educational systems, from elementary schools to universities, foreign investors create a win-win scenario: they can profit and eventually benefit from a skilled workforce, while local residents gain access to high-quality education at a lower cost. As a result, school enrollment rates in host countries rise, as people can now stay in their own countries while receiving world-class education.

Moreover, FDI inflows contribute to the diversification of educational options, providing more choices for individuals when planning their future careers. After completing university, some people may choose to pursue master's degrees, while others may opt to work for foreign companies, which often sponsor certifications related to their jobs instead of higher education.

**FDI reduces inequality**

The explanation for these impacts, as outlined in the paper by Im and McLaren (2015), is that inward FDI can compete with domestic capital for local workers, leading to a reduction in the income of domestic investors while increasing the income of domestic workers. This, in turn, helps narrow the income gap. This concept is drawn from the political argument of Pandya (2014), which suggests that ordinary people generally support FDI.

## 3.3 *Environment Quality*

### 3.3.1 Definitions and Domains

EQ has been also clarified by various organizations and scientists with different approaches. According to United Nations (1997), "Environmental quality is a state of environmental conditions in environmental media, expressed in terms of indicators or indices related to environmental quality standards." In this definition, what we could understand to compose EQ seems to be any kind of environmental conditions and not a specific state has been mentioned yet.

Meanwhile, one of widely known definitions of EQ mentioned by the European Environment Agency (2020) is "properties and characteristics of the environment, either generalized or local, as they impinge on human beings and other organisms. EQ is a general term which can refer to: varied characteristics such as air and water purity or pollution, noise, access to open space, and the visual effects of buildings, and the potential effects which such characteristics may have on physical and mental health (caused by human activities)." This definition of European Environment Agency (2020) specified any "properties and characteristics of the environment" which could have an impact (either positive or negative) on human beings or other organisms. In comparison with the previous definition of the United Nations (1997), this is regarded to be a more focused and detailed approach toward the effect on health (including both physical and mental health). In particular, "air and water purity or pollution, noise, access to open space, and the visual effects of buildings, and the potential effects" have been mentioned to be included in the EQ.

Relating to domains of EQ, to be more specific, the United States Environmental Protection Agency (2014) for the purposes of setting up the EQ index has summarized five domains: (1) air, (2) water, (3) land, (4) built, and (5) sociodemographic. This classification is based on the approach that these domains could significantly affect human health, which is aligned with what is concentrated on in the definition of European Environment Agency (2020).

### 3.3.2 Measurement

As mentioned by Anderson (2020), there have been many measures of EQ. These measures could be under the form of a composite index or particular measures (to look into specific areas of the environment).

Regarding the composite index, the United States Environmental Protection Agency (2014) did construct the EQ index on the basis of five domains: (1) air, (2) water, (3) land, (4) built, and (5) sociodemographic. In spite of being widely known, this index just considered the cases of more than 3000 counties in the United States. Besides this EQ index, other indices have been set up such as Environmental Sustainability Index or Environmental Vulnerability Index.

However, as it is not easy to fully and accurately measure the EQ via those composite indices across countries in the world, the other measures to look into a specific domain are preferable and popularly used. As the domain of air plays a very important role (normally ranked the first) to the environment, international organizations usually base on certain measures for this to present the situation of EQ. Some of those measures are:

- Total greenhouse gas (GHG) emission
- Carbon dioxide ($CO_2$) or carbon monoxide (CO) emission
- Nitrogen oxide (NO) or nitrogen dioxide ($NO_2$) emission

## 4 Theoretical Impacts of FDI on Environmental Quality

Regarding the impact of FDI on EQ, there exists contrasting points of view. While some research has found out the negative effect of FDI (meaning that the more FDI, the worse EQ), the other scientists support the significantly positive influence of this capital inflows on host countries. The theoretical effects of FDI on EQ could be clarified as follows:

### *4.1 Positive Impacts*

According to the OECD report (OECD, 2008), FDI can have a positive impact on the environment. It fosters new growth and structural efficiencies, enabling larger investments in environmental protection. Foreign investors may also introduce more advanced environmental protection technologies than those currently available in the host country.

The positive impact of FDI on EQ is primarily explained by the "Pollution Halos" hypothesis. According to Eskeland and Harrison (2003) "Moving to greener pastures? Multinationals and the pollution haven hypothesis," FDI-invested enterprises in four developing countries (Mexico, Venezuela, Morocco, and Côte d'Ivoire) were found to use energy more efficiently and adopt greener energy sources compared to domestic firms. Supporting the "Pollution Halos" hypothesis, Christmann and Taylor (2001), Blackman and Wu (1998), the Business and Industry Advisory Committee to the OECD (BIAC, 1999), and Merican et al., (2007) also argue that FDI facilitates the transfer of modern environmental technologies and management practices from developed to developing countries. As a result, host countries receiving FDI are seen as becoming "Pollution Halos."

## 4.2 Negative Impacts

In contrast to the "Pollution Halos" hypothesis, the "Pollution Havens" hypothesis highlights the negative impacts of FDI on EQ. In the paper "Foreign Direct Investment and the Environment: Pollution Haven Hypothesis Revisited", Aliyu (2005) explains the mechanisms behind these effects as follows:

At first, as carrying out FDI, multinational Companies (MNC) also relocated heavily polluting industries from home developed countries, which usually require the enforcement of stringent environmental policies, to host developing ones with just much looser and weaker conditions for such kinds of policies. This trend seems to be more popular in accordance with global free trade and deeper integration, which seriously harms EQ. This way of explanation has been also supported by Mani and Wheeler (1997) in the research of "*In Search of Pollution Havens? Dirty Industry in the World Economy, 1960-1995.*" Mani and Wheeler (1997) also indicated that despite the bad consequences to the environment, the amount of profit generated for FDI investors was extremely high, therefore, the value of FDI inflows to these "dirty" industries still accounted for a large proportion of the total one. This kind of finding has also been found in the research of Low and Yeats (1992).

Secondly, the negative effects on the environment of host countries also appear when the hazardous waste from developed countries (for instance from the process of industrial and nuclear energy production) is dumped in developing countries during the investment under the form of FDI. As a result, many host countries, which have received FDI, have been facing the threat of becoming a landfill. This raises an alarm to host countries.

Thirdly, in the competition for attracting more FDI, some countries can lower the environmental standards compared to other countries or they could resist increasing their standards, in order to gain a competitive advantage (OECD, 2008).

Lastly, an increase in FDI can have a notably negative impact on EQ due to the unchecked extraction of nonrenewable natural resources by investors motivated by resource-seeking. The unregulated and excessive exploitation of such resources (e.g., oil) can lead to rapid depletion. Esty and Gentry (1997) also highlighted that, of the three types of investor motives (market-seeking, resource/strategic assets-seeking, and efficiency-seeking), resource-seeking has the most significant adverse effect on the EQ of host countries. The greater the focus of investors on nonrenewable resources, the more detrimental the impact on the host country's environmental health.

# 5 Quantitative Analysis of the Impact of FDI on Environmental Quality

## *5.1 Model and Methodology*

To assess the impact of FDI on EQ, we used panel data and employed a model specification grounded in the theory of the environmental Kuznets curve (EKC) (further details on the environmental Kuznets curve can be found in Annex A1). The first equation is designed to examine the effect of FDI flows on $CO_2$ emissions (with greenhouse gas emissions used as a proxy for EQ in a robustness check). Next, we aim to investigate whether FDI has an inverted-U-shaped impact on EQ, as shown in Eq. (2).

$$LogCO2_{it} = \alpha_1 LogFDI_{it} + \alpha_2 LogGDPcap_{it} + \alpha_3 LogGDPcapsqt_{it} + \alpha_4 Open + \epsilon_{it} \quad (1)$$

$$LogCO2_{it} = \alpha_1 LogFDI_{it} + \alpha_2 LogFDIsqt + \alpha_3 LogGDPcap_{it} + \alpha_4 LogGDPcapsqt_{it} + \alpha_5 Open + \epsilon_{it} \quad (2)$$

Where: *i* denotes country *i* and *t* denotes year *t*.

- $\boldsymbol{LogCO2_{it}}$ is the logarithm of total $CO_2$ emissions in country *i* during year *t*
- $\boldsymbol{LogFDI_{it}}$ is the logarithm of the value of FDI inflows to country *i* during year *t*
- $\boldsymbol{LogFDIsqt_{it}}$ is the logarithm of the squared the value of FDI inflows to country *i* during year *t*
- $\boldsymbol{LogGDPcap_{it}}$ is the logarithm of the Gross Domestic Product (GDP) per capita of country *i* during year *t*;
- $\boldsymbol{LogGDPcapsqt_{it}}$ is the logarithm of the squared Gross Domestic Product (GDP) per capita of country *i* during year *t*;
- $\boldsymbol{Open_{it}}$ is to present the openness of country *i* during year *t* (This variable is calculated by taking the ratio of the total value of export and import and GDP of country *i* during year *t*).

The coefficient of interest in the model shown above are $\alpha_1$ measures the impact of FDI flows on the $CO_2$ emissions. If this capital flow actually increases the pollution level (over $CO_2$ emissions), the coefficient will be positive.

## *5.2 Database*

From official and trustworthy data sources, a sample has been consolidated and built to cover a time range of 1970 and 2019 for 230 countries in the world.

The data for all variables have been taken from official source of World Bank (https://databank.worldbank.org/home.aspx).

Summary statistics of variables are shown in Table 1:

## 5.3 Results and Interpretation

Table 2 presents the estimation results regarding the impact of FDI on EQ, with $CO_2$ emissions used as a proxy for EQ in Eqs. (1) and (2). The results are derived using panel data techniques, specifically fixed effects (FE) and random effects (RE).

For the relationship between FDI and $CO_2$ emissions, columns (1) to (4) show consistent findings across all variables studied. Regarding the ***LogFDI*** variable, the estimated coefficient is statistically significant at the 1% level and has a positive sign. This indicates that a 1% increase in FDI into Vietnam would lead to a rise in

**Table 1** Summary statistics of variables

| Variable | Obs | Mean | Std. Dev. | Min | Max |
|---|---|---|---|---|---|
| logco2 | 9578 | 9.72545 | 3.335725 | 1.299374 | 17.40286 |
| logghg | 9434 | 10.9435 | 3.294377 | 0.3294546 | 17.79568 |
| logfdi | 9714 | 19.87794 | 3.467521 | 2.302585 | 28.77401 |
| loggdpcap | 10,012 | 7.833412 | 1.614719 | 4.053333 | 12.15041 |
| open | 8826 | 0.7526036 | 0.5018487 | 0.0016742 | 8.608 |

Source: Authors' compilation

**Table 2** The estimation results of the impact of FDI on environmental quality (measured by $CO_2$ emission)

| | CO2 | | | |
|---|---|---|---|---|
| Variables | (1) | (2) | (3) | (4) |
| logfdi | 0.0425*** | 0.0761*** | 0.0516*** | 0.0597*** |
| | (0.00268) | (0.0105) | (0.00289) | (0.0114) |
| logfdisqt | | −0.000981*** | | −0.000232 |
| | | (0.000296) | | (0.000321) |
| loggdpcap | 1.706*** | 1.696*** | 1.684*** | 1.681*** |
| | (0.0299) | (0.0300) | (0.0323) | (0.0326) |
| loggdpcapsqt | −0.0845*** | −0.0832*** | −0.0839*** | −0.0836*** |
| | (0.00183) | (0.00187) | (0.00198) | (0.00204) |
| open | 0.207*** | 0.219*** | 0.160*** | 0.161*** |
| | (0.0188) | (0.0191) | (0.0203) | (0.0206) |
| Observations | 7396 | 7396 | 7396 | 7396 |
| Number of id | 230 | 230 | 230 | 230 |
| Type | FE | FE | RE | RE |

Estimation obtained by using Panel data techniques of fixed and random effect. ***/**/* denote the significance level of 1%/5%/10% of the t-statistic

Source: Authors' compilation

$CO_2$ emissions by approximately 0.04% to 0.08%. Therefore, it can be concluded that FDI flows into 230 countries worldwide, through investments in production facilities and business activities, have contributed to an increase in $CO_2$ emissions.

In order to see if there exists the inverted-U shape of FDI on EQ, as looking into the coefficients of ***LogFDIsqt***, we could see that for fixed-effect estimation (which has an advantage of controlling for time-invariant characteristics of countries), despite having the not so high magnitude, the coefficient is statistically significant and negative. This could partly prove that early FDI will highly lead to the rise in $CO_2$ emissions; however, after reaching a certain value, the impact gets reversed to some extent. This could be a good signal for host countries which receive FDI all over the world.

Regarding ***logGDPcap*** and ***logGDPcapsqt***, their statistical significance at the 1% level, along with the signs of the estimated coefficients, clearly illustrates the impact of income on the EQ of host countries, as described in the environmental Kuznets curve theory. Specifically, the logarithm of GDP per capita (***logGDPcap***) has a positive sign, while the square of GDP per capita (logGDPcapsqt) has a negative sign. This indicates the presence of an inverted-U relationship between national income and $CO_2$ emissions. For ***logGDPcap***, the coefficient is approximately 1.7, meaning that a 1% increase in GDP results in a 1.7% rise in emissions. The magnitude of this coefficient for logGDPcap is considerably higher than that for FDI, suggesting that changes in a country's income have a greater impact on the environment than FDI inflows. This aligns with real-world observations.

Regarding the variable ***Open*** (the ratio of the sum of exports and imports to GDP), we found that all coefficients are positive and statistically significant. These results reflect the reality that greater openness to trade in countries tends to lead to higher $CO_2$ emissions. The magnitudes of the coefficients range from 0.16 to 0.22, which are higher than those of LogFDI but lower than those of LogGDPcap.

**Robustness check**

For robustness check, we use the greenhouse gas (GHG) emission as another proxy for EQ. From Table 3, we have also found the consistent results, especially the significantly positive impact of FDI on GHG emission, meaning the more FDI, the worse EQ of host countries is. In addition, the effect of national income on EQ is also statistically significant and environmental Kuznets curve theory has also been proved. Furthermore, the openness of countries to trade also makes it worse for the EQ with more GHG emissions.

## 6 Conclusions

Through theoretical analysis, we have identified clear evidence supporting the negative impact of FDI on EQ. This indicates that FDI inflows into countries contribute to deteriorating environmental conditions. These findings validate the "pollution haven" hypothesis. However, this effect typically occurs during the early stages of

**Table 3** The estimation results of the impact of FDI on environmental quality (measured by GHG emission)

| | ghg | | | |
|---|---|---|---|---|
| Variables | (1) | (2) | (3) | (4) |
| logfdi | 0.0377*** | 0.0433*** | 0.0508*** | 0.0226* |
| | (0.00293) | (0.0110) | (0.00318) | (0.0120) |
| logfdisqt | | −0.000166 | | 0.000843** |
| | | (0.000314) | | (0.000342) |
| loggdpcap | 0.986*** | 0.984*** | 0.948*** | 0.959*** |
| | (0.0337) | (0.0339) | (0.0368) | (0.0372) |
| loggdpcapsqt | −0.0493*** | −0.0490*** | −0.0483*** | −0.0497*** |
| | (0.00207) | (0.00214) | (0.00227) | (0.00235) |
| open | 0.133*** | 0.135*** | 0.0598** | 0.0463* |
| | (0.0225) | (0.0230) | (0.0244) | (0.0249) |
| Observations | 6783 | 6783 | 6783 | 6783 |
| Number of id | 220 | 220 | 220 | 220 |
| Type | FE | FE | RE | RE |

Estimation obtained by using Panel data techniques of Fixed and Random effect. ***/**/* denote the significance level of 1%/5%/10% of the t-statistic
Source: Authors' compilation

FDI inflow. Once a certain threshold is reached, higher levels of FDI may lead to improvements in EQ. Nonetheless, this result warrants further verification.

In reality, the improvements of EQ do not come automatically together with the economic growth but they are dependent on the policy of host countries. Host countries, especially less-developed countries, have to decide to change the way they attract and choose FDI: foreign investors with high-standard on the environment have to be welcomed.

Additionally, the study highlights other key findings, such as (i) an inverted U-shaped relationship between pollution levels and national income, as proposed by the environmental Kuznets curve theory and (ii) greater trade openness in countries leads to a decline in EQ, as both CO2 and GHG emissions increase.

## Appendix 1: The Environmental Kuznets Curve

The Kuznets curve was introduced by Simon Kuznets at the annual meeting of the 67th American Economic Association in December 1954. By 1991, this curve had become one of the important means to describe the relationship between environmental quality and per capita income over time. According to the theory of Kuznets Curve, the level of environmental degradation and per capita income follow the rules of the inverted U-shape, i.e.: in the early stages of development, environmental

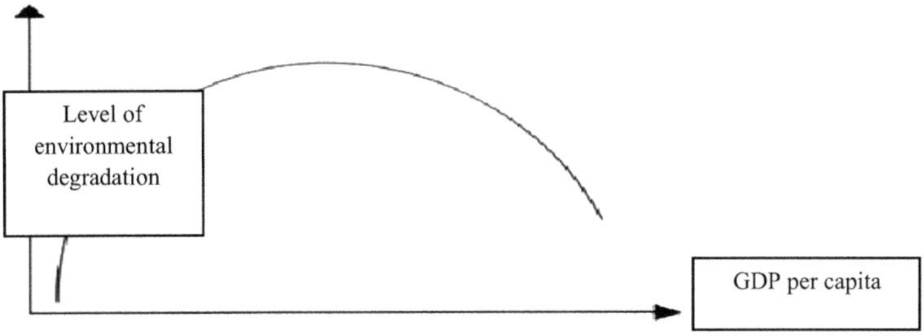


**Fig. 2** The environmental Kuznets Curve

degradation increases and peaks at the top, then when per capita income exceeds a specific threshold, the degradation will diminish (Fig. 2).

There are some specific explanations for the increase or decrease in the degree of degradation as follows:

Firstly, *in the early stages*, especially in the beginning of the process of industrialization, countries often prioritize production activities, trading, and increasing output capacity to help people increase their income. This leads to the use of natural resources followed up by emissions into the environment in large quantities, which consequently causes environmental pollution to increase rapidly.

However, when entering *the later stages of the process of industrialization*, people's incomes are higher and they are more aware of the value of environmental protection. Besides, policies and laws on the environment of the state agencies have become more stringent. This leads to the widespread application of clean technologies, producing advanced and environmentally friendly products, thus helping to reduce pollution.

## Appendix 2: The Correlation among Variables

| e(V) | logfdi | loggdpcap | log2gdpcap | open | _cons |
|---|---|---|---|---|---|
| logfdi | 1 | | | | |
| loggdpcap | −0.2326 | 1 | | | |
| log2gdpcap | 0.0634 | −0.9711 | 1 | | |
| open | −0.1731 | 0.0368 | −0.0679 | 1 | |
| _cons | −0.0434 | −0.9474 | 0.9438 | −0.0452 | 1 |

These correlations are associated with the regression of the Eq. (1), column (1)

## References


Aliyu MA (2005) Foreign direct investment and the environment: pollution haven hypothesis revisited. Eighth annual conference on global economic analysis, Lübeck

Anderson TL (2020) Environmental quality. Econlib. Retrieved from https://www.econlib.org/library/Enc/EnvironmentalQuality.html

BIAC (1999) BIAC discussion paper on FDI and the environment. OECD conference on foreign direct investment and the environment, The Hague

Blackman A, Wu X (1998) Foreign direct investment in China's power sector: trends, benefits and barriers. *Energy Policy* 27(12):695–711. https://doi.org/10.1016/S0301-4215(99)00063-4

Christmann P, Taylor G (2001) Globalization and the environment: determinants of firm self-regulation in China. J Int Bus Stud 32(3):439–458

Dinh HL, Lin SM (2014) CO2 emissions, energy consumption, economic growth and FDI in Vietnam. Managing Global Transition 12(3):219–232

Eskeland SG, Harrison AE (2003) Moving to greener pastures? Multinationals and the pollution haven hypothesis. J Dev Econ 70:1–23

Esty D, Gentry B (1997) Foreign investment, globalisation and environment. Globalization and the Environment:141–172

European Environment Agency (2020) Environmental quality. https://www.eea.europa.eu/help/glossary/eea-glossary/environmental-quality

Hitam MB, Borhan HB (2012) FDI, growth and the environment: impact on quality of life in Malaysia. Procedia Soc Behav Sci 50:333–342. https://doi.org/10.1016/j.sbspro.2012.08.038

Im H, McLaren J (2015) Does Foreign Direct Investment raise income inequality in developing countries? A new Instrumental variables approach

International Monetary Fund (IMF) (1993) Balance of payment, 5th edition. https://www.imf.org/external/np/sta/bop/BOPman.pdf

Karlsson S, Ludin N, Sjoeholm F, He P (2007) FDI and job creation in China. IFN Working Paper. 273. Available at: https://www.econstor.eu/bitstream/10419/81217/1/wp723.pdf

Kirkulak B, Qiu B, Yin W (2011) The impact of FDI on air quality: evidence from China. J Chin Econ Foreign Trade Stud 4(2):81–98(18)

Kurtishi-Kastrati S (2013) The effects of foreign direct investments for host country's economy. Eur J Interdiscip Stud 5:26–38

Lee CG (2009) Foreign Direct Investment, pollution and economic growth: evidence from Malaysia. Appl Econ 41:1709–1716

Low P, Yeats A (1992) Do "dirty" industries migrate?. World Bank Discussion Papers

Mani M, Wheeler D (1997) In search of pollution havens? Dirty industry in the world economy, 1960–1995. Presentation to the OECD conference on FDI and the environment (The Hague, 28–29 January 1999), 7(3). https://doi.org/10.1177/107049659800700302

Merican Y, Yusop Z, Noor ZN, Law SH (2007) Foreign direct investment and pollution in five ASEAN nations. Int J Econ Manag 1(2):245–261. https://www.ijem.upm.edu.my/vol1no2/4.pdf

OECD (2008) Benchmark definition of Foreign direct investment, 4th edition 2008. https://www.oecd.org/investment/fdibenchmarkdefinition.htm

Pandya SS (2014) Democratization and foreign direct investment liberalization, 1970–2000. Int Stud Q 58(3):475–488

Pao H, Tsai CH (2011) Multivariate granger causality between CO2 emissions, energy consumption, FDI (Foreign Direct Investment) and GDP (gross domestic product): evidence form a panel of a BRIC (Brazil, Russian Federation, India, and China) countries. Energy 36:685–693

The United States Environmental Protection Agency (2014) Environmental Quality Index - Overview report. https://edg.epa.gov/EPADataCommons/public/ORD/NHEERL/EQI/EQI%20Overview%20Report__Final.pdf

To AH, Ha DT, Nguyen HM, Vo DH (2019) The impact of Foreign Direct Investment on environment degradation: evidence from emerging markets in Asia. Int J Environ Res Public Health 16:1636. https://doi.org/10.3390/ijerph16091636

Tran TH (2005) Impacts of Foreign Direct investment on poverty reduction in Vietnam. Discussed paper, GRIPS, https://gdforum.sakura.ne.jp/VDFTokyo/Doc/18TTHungPaper.pdf
UNCTAD (1998) World investment report 1998: trends and determinants overview. https://unctad.org/system/files/official-document/wir1998overview_en.pdf
United Nations (1997) Glossary of environment statistics, studies in methods, Series F, No. 67, New York
Vu CL (2012) International investment. Lao dong Publishing House

**Vinh-Thi Hong Cao** has been granted her PhD degree in International Economics at Foreign Trade University, Vietnam and Master in Economics at Mannheim University, Germany. She has been working for School of Economics and International Business, Foreign Trade University. Her research interests are international investment, international trade, innovation, firm performance, learning organization, etc.

**Dung-Tien Tran** is currently working for Urban Development Agency, Ministry of Construction, Vietnam. He completed his PhD at the University of Lyon and graduated in civil and construction engineering from INSA de Lyon. His research focuses on such topics as public investment, sustainable development, urban planning, and urban economics.

# Environmental Provisions Within New-Generation FTAs: Motivations, Implications, and Challenges

**Ngan-Kim Vu and Hang-Thi Minh Hoang**

**Abstract** To date, Vietnam has been a party to new-generation FTAs, including the Comprehensive and Progressive Agreement on Trans-Pacific Partnership (CPTPP) and the EU-Vietnam Free Trade Agreement (EVFTA). The two agreements share certain objectives, ranging from the enhancement of the mutual supportiveness between trade policies and environment policies to the high level of environmental protection as well as the effective implementation of national environment laws, capacity of parties in the settlement of trade-related environment issues, and environment management protection. Meanwhile, since the turn of the 1990s, domestically Vietnam has set stepping stones for its sustainable development pursuit whereby environment protection is an inextricable part of the sustainable development strategy—the inevitable development roadmap of Vietnam. This chapter contends, *firstly*, the compatibility between the environmental objectives of the two agreements and those in Vietnam's sustainable development policies. In particular, the accomplishment of environment-related objectives under these FTAs also serves the goals of efficient exploitation and utilization of natural resources of the strategy on sustainable development in Vietnam. *Secondly*, through analyzing environment-related provisions within Vietnam's new-generation FTAs such as CPTPP and EVFTA, the research tackles the implications as well as challenges from legal and institutional perspective, and consequently provides some recommendations for Vietnam in the effective implementation of these provisions. Undoubtedly, to further promote sustainable development and particularly environment-related matters, it requires not only the authority, agencies, enterprises but the whole community of Vietnam to join forces.

N.-K. Vu (✉) · H.-T. Minh Hoang
Foreign Trade University, Ha Noi, Vietnam
e-mail: nganvk@ftu.edu.vn; hanghtm@ftu.edu.vn



H.-T. Vu, L.-P. Dana (eds.), *Vietnam on the Road to Sustainable Economic, Social and Business Development*, Frontiers in South and Southeast Asian Development Research, https://doi.org/10.1007/978-981-96-4371-4_6

# 1 Introduction

As a low middle-income developing country, Vietnam is still at the very beginning of its development process featuring growth policies detrimental to the environment. More specifically, according to the Kuznets environment curve, which depicts the relationship between development and environment as an inverted U shape, at the beginning of the development process, states pay more attention to the economic growth concern at the expense of the environment. In other words, as the economy starts to develop, the environment has the tendency to become worse. At the later stage of development when basic needs of the people are met, a more developed economy is followed by better environment. In light of the status quo of the economic development in Vietnam, it is arguable that its economic development is in tension with the environment protection mandate. Given the labor-and resource-intensive competitiveness on which the economic structure of Vietnam is largely based, the expansion of production activities even exacerbates the overexploitation of natural resources to the detriment of the environment. Moreover, increased incomes following economic development lead to more consumption, putting more pressures on the environment due to wastes therefrom. Additionally, due to limited financial resources allocated to environment policies, clean and modern technologies are unaffordable to relevant parties in the economy which in turn worsens the environment quality. Last but not least, both the importation of wastes and exportation of endangered species are foes to the environment.

However, in recent years, Vietnam has formally informed the adoption of a swift in the development direction/strategy toward sustainable development (presented in several documents including *the Vietnam's Socioeconomic Development Strategy for 2011–2020, the National Green Growth Strategy 2012, the National Strategy on Environmental Protection to 2020 with vision to 2030, and the Law on Environmental Protection 2014*). Sustainable development is the development that meets the needs of the current generation without undermining the ability of future generations to satisfy their own demands. Meanwhile, from 2000 onward, Vietnam has actively negotiated and concluded new-generation free trade agreements (FTAs) which explicitly recognize environment protection, among others, as one of the main objectives in their preamble in addition to ranges of relevant substantive provisions. In particular, the Comprehensive and Progressive Agreement for Trans-Pacific Partnership (CPTPP) concluded in March 2018 covers Chap. 20 on Environment while the EU-Vietnam Free Trade Agreement (EVFTA) concluded in June 2018 contains Chap. 13 on Trade and Sustainable Development. The ensuing question is how Vietnam can materialize such a nationwide development policy change in favor of environment protection when its FTA-included environment rules are concerned. To put in another way, the issue at question is whether and how environmental provisions in new-generation FTAs contribute to the achievement of the pro-environment policy switch in Vietnam.

This chapter aims at elaborating the consistency of environmental provisions in new-generation FTAs with the domestic objective of sustainable development in

Vietnam, analyzing commitments generated under these agreements and providing some recommendations for Vietnam in the implementation process. It is structured in three sections. Section 1 analyzes whether and how these environmental commitments are in line with the sustainability objective as set out by the Vietnamese government. Section 2 then analyzes environmental obligations derived from FTAs, and concerns the impacts of these provisions on relevant parties ranging from the authorities to enterprises and the society. Section 3 concludes with some recommendations for Vietnam to better take advantage of these regulations.

## 2 Methodology

The chapter uses desk research, doctrinal research method, socio-legal analysis and case synthesis. A doctrinal approach, that is, "black letter" legal research method, typically focuses on statutes, case law, and other legal sources. It attempts to look at the law within itself and thus, it examines law as a written body of principles which can be discerned and analyzed using only legal sources. At the same time, socio-legal methodology is employed to evaluate impacts of these FTA environmental provisions towards relevant groups/parties in Vietnam.

Opinions of experts and scholars may be utilized in order to shed the light on the understanding of impacts of FTA environmental provisions as well as to recommend possible solutions.

## 3 Key Findings

- The analysis of environmental provisions within new-generation FTAs of Vietnam and the sustainable development strategy in Vietnam showcases the consistency inter se on paper
- Analysis of FTA environmental provisions, identifying challenges from legal and institutional perspectives and provide some policy implications on relevant groups: the authority, enterprises, consumers, and the society as a whole
- Recommendation for Vietnam in the implementation of FTA commitments relating to the protection of the environment

Even though the environmental provisions within FTAs of Vietnam are somehow in line with relevant laws, regulations, and policies of Vietnam, these turn out to be demanding and thus, lots of difficulties are anticipated in the implementation of such provisions. In order to tackle these issues, efforts should be made, inter alia, to revise or modify relevant laws and regulations in order to improve the legal framework for environment protection in Vietnam, establish appropriate policies to encourage enterprises to take initiatives relating to sustainable development, and increase awareness of consumers/society in the protection of the environment.

# 4 Discussion

## 4.1 *The Consistency of Objectives Between Environment Provisions in New-Generation FTAs and the Sustainable Development Strategy of Vietnam*

Regardless of the absence of a single, unanimously agreed understanding of new-generation FTAs, they share some common features. Being often subsequent to the WTO establishment in 1995 and concluded by signatories more limited in number than those in multilateral trade agreements, trade agreements belonging to the so-called new-generation feature deeper and wider commitments in trade and investment compared to the WTO law. So far, Vietnam has been a party to a number of such FTAs, including, notably, the Comprehensive and Progressive Agreement on Trans-Pacific Partnership (CPTPP) and the EU-Vietnam Free Trade Agreement (EVFTA).

Since the turn of the 1990s, domestically Vietnam has set stepping stones for its sustainable development pursuit whereby environment protection is an inextricable part of the sustainable development strategy—the inevitable development roadmap of Vietnam. Meanwhile, only since the CPTPP which was negotiated in the 2010s, Vietnam has internationally undertaken explicit environment commitments within an FTA setting. Subsequently, along the same line, Vietnam agreed on environment provisions in the EVFTA. The lag in the time points between these frameworks gives rise to concerns about the consistency inter se, or, to put it in another way, whether objectives of provisions in Environment Chapter in the CPTPP as well as the equivalence in the EVFTA are consonant with those of the national sustainable development strategy in Vietnam. This part sheds light on the compatibility of objectives between these two frameworks. More specifically, it evaluates whether and to what extent they are consistent with each other. It first introduces the sustainable development strategy in Vietnam. Objectives of environment provisions in the two FTAs are elaborated in detail afterward. This part concludes with assessment of the consistency therebetween.

## 4.2 *Sustainable Development Strategy in Vietnam*

Internationally, the notion of sustainable development was first introduced in the 1970s; in the meantime, in Vietnam, the National Plan on Environment and Sustainable Development during the 1991–2000 period stipulated in Decision 187-CT dated 12/6/1991 was the precursor for sustainable development policies in Vietnam. The Law on Environment Protection was enacted for the first time in 1993. Afterwards, Directive 36/1998/CT-TW dated 25/6/1998 issued by the Political Bureau on the enhancement of environment protection activities during industrialization and modernization highlighted that environment protection is an essentially inevitable component of the socioeconomic development roadmap, guidelines, policies, and plans at all levels and sectors, and constitutes a crucial foundation for sustainable development as well as the successful realization of the industrialization and modernization cause.

Documents at the IX Grand Meetings of Nationwide Delegates of the Communist Party in Vietnam and the Socioeconomic Development Strategy in the 2001–2010 period persistently reaffirmed the perspective of Vietnam on sustainable development. In 2004, Decision 153/2004/QĐ-TTg was issued by the Prime Minister of Vietnam on the adoption of the Orientations toward Sustainable Development in Vietnam, or the so-called Agenda 21 of Vietnam, based on which the Law on Environment Protection 2005 was enacted. Thereafter, in 2012, the then Prime Minister issued the Decision 432/QĐ-TTg on the strategy of sustainable development of Vietnam in the 2011–2020 period. The Decision specifies that, in terms of perspectives, sustainable development, according to the stance of Vietnam, involves three components: the economy (economic development), society (social development), and environment (protection of resources and the environment). In terms of objectives, overall objectives are conservation of natural resources and the environment, in addition to the equivalent with regard to economic growth and social development. Likewise, besides concrete objectives relevant to economic growth and social development, the like, when it comes to the protection of resources and the environment, entail: (i) minimization of adverse impacts of economic activities on the environment; (ii) efficient exploitation and usage of natural resources, especially exhaustible ones; and (iii) prevention, surveillance, and correction of pollutions and biodiversity preservation. The sustainable development strategy, although failing to make a direct reference to the trade-vis-a-vis-environment nexus, does so implicitly through its mention of the general interaction between the economy and the environment. Subsequently, the Law on Environment Protection 2014 was enacted. Although being in a multiple quantity, these documents generally speaking adopt a common international understanding of sustainable development, that is, "the development which meets the present needs without circumventing the satisfaction of the needs of future generations" (Article 3.4, Law on Environment Protection 2005; Article 3.4, Law on Environment Protection 2014) and cover three facets: economic growth, social development, and environment protection (van Aaken & Lehmann, 2011).

## *4.3 Objectives of Environment Provisions in the CPTPP and the EVFTA*

Besides two annexes, environmental provisions stipulated in Chap. 20 on Environment of the CPTPP total 23. Article 20.2 of the CPTPP elaborates seven objectives of the chapter and their mutual supportiveness. Meanwhile, in the EVFTA, environment provisions are stipulated in Chap. 13 on Trade and Sustainable

Development and count 17 with five main objectives. The two agreements share certain objectives regarding the environment.

*First,* their environmental provisions aim at the enhancement of the mutual supportiveness between trade policies and environment policies (Article 20.2.1 of the CPTPP; Article 13.5 of the EVFTA). The two sets of policies can interact with each other in either "win-win"—mutual supportiveness or "win-lose"—conflict scenarios. Their mutual supportiveness denotes situations in which trade policies foster environment protection and vice versa environment policies enhance free trade. Their mutual supportiveness may arise in certain circumstances when this pair of policies is in pursuit of common goals (Bodansky & Lawrence, 2009). Trade policies prohibiting agricultural or fishery subsidies to the overexploitation of natural resources, for instance, are in line with environment protection goals of environment policies. Nevertheless, trade policies may be hostile to environment policies in certain conditions where their goals are in contrast. For instance, trade policies target at the removal of barriers to free trade, which in turn, in certain contexts, are instrumental tools in environment policies to correct market failures. As a matter of fact, it is understandable that the first objective of the two-FTA environment regulations is to foster mutual supportiveness, or limit the conflict inter se, which in turn can be translated into the assurance of the accordance between objectives of the two policies.

*Second*, their environment provisions are also designed for the purpose of fostering the high level of environment protection as well as the effective implementation of national environment laws (Article 20.2.1 of the CPTPP, Article 13.2.2 of the EVFTA). With regards to newly proposed, higher-standard but not-yet-adopted environment regulations, FTA parties, including Vietnam, have undertaken commitments to neither resistance nor withdrawal of their adoption. Likewise, with respect to already adopted environment provisions, the parties make promises not to repeal, or fail to implement, or ineffectively implement. This objective has its root in the negative influence of environment policies on trade policies. At the micro level, confronting with highly demanding environment policies in one nation which entails high compliance costs, goods producers or services providers have to cope therewith to remain competitive by various means, including reduction of environment compliance costs through, inter alia, the relocation of their production facilities to other nations with less-demanding environment standards, thereby saving expenses on environment-regulation compliance (Bodansky & Lawrence, 2009). At the macro level, in view of the abovementioned defensive responses from the private actors in the economy, state regulators or nations in general, in order to remain competitive in their trade and investment attraction, may adopt the strategy of "a regulatory chill" and/or "a race to bottom" (Ibid). In "a regulatory chill", state regulators may neither adopt any modern and higher-standard environment requirements nor effectively implement adopted environment standards or regulations. In "a race to bottom," nations even lower their existing, concurrently implemented environment standards.

*Thirdly*, another objective is the enhancement of capacity of parties in the settlement of trade-related environment issues, including through cooperation (Article

20.2.1 of the CPTPP, Article 13.1.4 of the EVFTA). Trade-related environment issues cover two main groups: trade-related environment measures (TREMs) to the contrary of the free trade objective, that is, environment measures equivalent to non-tariff barriers, and TREMs to the contrary of the fair trade objective, that is, measures that bring out proper competitiveness. Over the past few years, within the framework of the WTO, there have been multiple disputes over TREMs which either constitute non-tariff barriers or yield unfair competitiveness. Nations adopting TREMs routinely justify their measures as legitimate in state sovereignty according to which it is the discretion of states to adopt and enforce national environment laws in pursuit of public policies, objectives, or order. In contradiction, parties on the other side often counterargue over adverse impacts of these measures on free trade and/or fair trade. By incorporating this objective, parties predict potentials for similar disputes inter se and commit to different channels for dispute resolutions, including negotiation and collaboration together with the dispute settlement mechanism in the WTO.

The fourth objective is the contribution to sustainable development (Article 20.2.2 of the CPTPP, Article 13.1.1 of the EVFTA). Neither of the two FTAs defines the sustainable development concept; however, the EVFTA manifests its three components, including economic development, social development, and environmental protection, as well as their nexus of interdependence and mutual reinforcement (Article 13.1.1 of the EVFTA). It is arguable that the first two objectives facilitate the achievement of this one. However, this objective takes into account concrete priorities or circumstances of each nation given the participation of both developed and developing trade partners (Article 20.2.2 of the CPTPP). As to developed countries, thanks to the highly-developed status quo of the economies, the more the economic growth, the less the pollution due to income increases, the availability of financial resources and the affordability of clean and modern environment technologies (Charnovitz, 2013). Domestically, in light of the needs of citizens for improved living quality, these countries have enacted statutes reaffirming the environment management objective in particular and sustainable development objective in general. Politically, domestic interest groups including nongovernmental organizations, civil societies, and environmentalists have raised concerns over the insertion of environment issues into FTA to put down adverse implications of free trade on the environment. On the other side, developing countries are at the very beginning of their economic development; therefore, the more grown the economy, the worse the environment pollution (Ibid). Furthermore, these nations just afford bare resources for the purpose of environment protection and often give priorities to economic development goals over other environment correspondence. Consequentially, their environment systems feature lower environment standards and ineffective environment law implementation and enforcement mechanisms. Voices of interest groups for environment protection are often unheard.

The fifth objective is to promote environment management (Article 20.2.2 of the CPTPP). As to developing countries like Vietnam, environment management capacity of different actors in the economy in general and of the state in particular is poor by virtue of factors such as sciences, technologies, know-how, finance, and humans detrimental to the enforcement of environment policies and environment quality. Hence, to obtain this objective, developing countries on their own and/or in collaboration with other developed partners through technical assistance to improve their competence in environment management in support of the achievement of other objectives.

Last but not least, the inappropriateness of conducts of FTA parties in establishment or utilization of environmental laws and other instruments to disguise restrictions on trade and environment is ascertained by both agreements (Article 20.2.3 of the CPTPP. Article 13.3.4 of the EVFTA). This objective mainly comes from the legal practice in developed, instead of developing, countries. In particular, developed countries impose either overexcessively demanding or redundant domestic environmental standards and regulations on export goods (producers) or services (suppliers) or investment originating in developing countries. Failing to meet these requirements will block their access to developed countries; otherwise, when satisfying these requirements, their exporting prices are no longer competitive in the importing market. These measures constitute disguised nontariff barriers to trade and investment to illegally protect domestic industries and employments which go against the trade and investment objectives of the FTAs. However, regardless of the concealing by developed countries of their underlying and real purposes by referring to other allegedly legitimate objectives such as public policy, objectives or order, these measures are increasingly subject to claims from developing countries.

## 4.4 Compatibility of Objectives Between FTA Environment Provisions and the Sustainable Development Strategy of Vietnam

Although, trade and environment policies can theoretically interact with each other according to two opposite scenarios (Bodansky & Lawrence, 2009), given the practical conditions in Vietnam, these two policies are very likely to find themselves in the contradictory circumstances. As above analyzed, being at the beginning of the economic development, the more Vietnam develops, the worse the environment pollution turns for (Fickling & Hufbauer, 2012). Additionally, objectives of the free trade policy in Vietnam in certain cases may be detrimental to the environment quality, that is, against the environment policy objectives. In particular, free trade in Vietnam can do harm to the environment through five different channels: production changes (i.e., domestic production expansions ensuing from free trade triggers wider scales of natural resources exploitation which in turn leads to environment degradation), income changes (increased incomes due to free trade produce more consumption needs which then put further pressures on the environment), structural changes (in light of competitiveness of Vietnam in natural-resource-extensive sectors, the blooming of these sectors follows free trade,) and physical changes (for instance, through importation of wastes or exportation of natural resources and overexploited or endangered animals and plants) (Charnovitz, 2013).

In light of this circumstance, environmental provisions in new-generation FTAs are compatible with sustainable development objectives of Vietnam. As abovesaid, there can be various means to sustainable development; however, within the FTA framework, the objective can be arrived at through the assurance of the mutual supportiveness between the trade policy and environment policy. In accordance with their objective, Vietnam is obligated to make its trade policy supportive of its environment counterpart and vice versa the latter in line with the former. The assurance of mutual supportiveness of policies on trade and environment will make contribution to the mitigation of hostile effects of economic activities on the environment protection in accordance with the sustainable development strategy. Furthermore, that the two policies support each other entails their shared goals of effective usages of resources in the economy. Therefore, the accomplishment of this objective under the new-generation FTAs also serves the goals of efficient exploitation and utilization of natural resources of the strategy on sustainable development.

Empirical studies have proven that impacts of trade liberalization on the environment are normally through indirect channels and contingent on the presence of an effective environment policy in place in the trade-liberalizing party (OECD, 2001). As a developing country, Vietnam has its environment regulations and standards to be featured by loose implementation and enforcement. However, once being bound to the commitment of mutually supportiveness of the two policies, Vietnam must take pains in adoption, implementation and enforcement of an effective environment policy. To put it differently, to accomplish this objective, Vietnam is required to adopt, implement, and enforce in practice an environment policy of two attributes: (1) strict accordance with the "polluter pays" principle, that is, having polluters to cover in full costs that the society incurs, and (2) no undertaking of environment measures distortive to the market or trade, such as subsidies to environment-hostile goods production or service provision activities (Ibid). Moreover, Vietnam is refrained from maintenance of low environment standards and/or ineffective implementation or enforcement of already adopted environment regulations domestically so as to remain price-competitive in trade due to environment-compliance cost-saving. This objective will contribute to the completion of all three concrete goals of resource and environment protection in the environment protection strategy.

## 4.5 *Analysis of Environment-Related Provisions Within New-Generation FTAs of Vietnam: Challenges and Implications from Legal and Institutional Perspectives*

As previously mentioned, new-generation FTAs refer to trade agreements that not only generate deeper level of commitments but could also go beyond traditional topics, for example, tariff and nontariff barriers for trade in goods and trade in services, to include other trade issues that are not currently under the framework of the WTO such as competition and e-commerce, or even nontrade matters such as labor and environment. Of these, the protection and preservation of the environment is the topic that has received much attention from FTA signatories and/or parties and has been included in an increasing number of new-generation FTAs (Hofmann et al., 2017). These environment-related provisions could be found in a chapter on sustainable development as in, for example, the EVFTA; or be remarkably put under a dedicated chapter on environment such as Chap. 20 of the CPTPP. Hence, the depth and breadth of these commitments vary from agreement to agreement while some topics have received more attention than others. Level of commitments generally reflects parties' interest and concerns in balancing, on the one hand, their aims to promote trade and, on the other hand, their desire to protect public welfare objectives toward sustainable development which include the protection and preservation of the environment (Preamble of the CPTPP). Even though Vietnam has been a party to many other new-generation FTAs such as the Vietnam–Republic of Korea FTA, the Vietnam–Eurasian Economic Union FTA (Eurasian Economic Union (EAEU) includes Russia, Armenia, Belarus, Kazakhstan and Kyrgyzstan) or the Regional Comprehensive Economic Partnership (RCEP) where Vietnam participates as an ASEAN Member, CPTPP and EVFTA remain the country's most ambitious agreements with respect to commitments on environmental issues. The following analysis thereby focuses on these two agreements as especially respect to the chapter on environment under the CPTPP.

## 4.6 Obligations Generated by Environment-Related FTA Provisions

Despite being included in a trade agreement, sustainable development in general or particularly environment is in fact non-trade issues. The inclusion of non-trade issues in a trade agreement proves efforts of contracting parties to ensure that the promotion of their economic, investment, and trade relationships shall be "in accordance with the objective of sustainable development" as stated in the preambles of both CPTPP and EVFTA.

This objective is then reinforced in Chap. 20 of the CPTPP, a dedicated chapter to environment, specifically to "*promote mutually supportive trade and environmental policies; promote high levels of environmental protection and effective enforcement of environmental laws; and enhance the capacities of the Parties to address trade-related environmental issues, including through cooperation" (Article 20.2 of the CPTPP).* In such a way, first and foremost, CPTPP parties recognize the importance of "mutually supportive trade and environmental policies and practices to improve environmental protection in the furtherance of sustainable development," while being mindful that each member country has its own sovereign right to establish, adopt, or modify its environmental laws and policies in the implementation of environment-related obligations under this agreement (Article 20.3(1) & (2) of the CPTPP). Admittedly, CPTPP does not require parties to enact or enforce new environmental law but leave the matter for each party to decide at its discretion as long as relevant domestic law and policy are in line with commitments made therein.

Generally speaking, Chap. 20 on Environment of the CPTPP provides for environment-related commitments which could actually cover a full spectrum of legal bindingness, from legally binding to nonbinding commitments. Binding commitment generates hard obligation and nonbinding commitment tends to render rather soft obligation. While implementation of the latter could hardly be subject to dispute settlement, implementation of the former could well be filed a dispute against it.

At one end of the spectrum, CPTPP does not require parties to participate in new multilateral environmental agreements (MEAs), but rather each party is to recognize the importance of MEAs and affirms to implement its respective commitments under the MEAs to which it is a party (Article 20.4 of the CPTPP). Specifically, for Vietnam, the country is required to, inter alia, implement and comply with provisions of the Montreal Protocol on Substances that Deplete the Ozone Layer (1987) to control the production, consumption of and trade in substances that can significantly deplete or modify the ozone layer (Article 20.5 of the CPTPP); take measures to prevent pollution of the marine environment from ship in accordance with the MARPOL protocol (Article 20.6 of the CPTPP); and fulfill its obligations under the Convention on International Trade in Endangered Species of Wild Fauna and Flora (CITES) to combat illegal trade in those natural resources (Article 20.17 of the CPTPP). Since Vietnam has already joined these agreements, admittedly, the above-mentioned CPTPP provisions do not generate additional obligations for Vietnam but to reinforce its existing obligations under MEAs in recognition of the importance of proper and effective implementation of multilateral environmental agreements. In such a way, the reaffirmation of existing commitments made under other agreements, that is, MEAs, is actually "not commitment in and of itself and thus, it cannot be binding." (UN Environment, 2016).

When it comes to specific issues of environment protection, commitments are mostly couched in best endeavor language. For instance, Article 20.13 of the CPTPP requires its parties to "promote" and "encourage" the conservation and sustainable use of biological diversity in accordance with its law or policy. According to Article 20.17, the parties commit to also "promote" conservation and combat illegal trade in wild fauna and flora through exchange relevant information and experiences and/or "endeavor" to implement CITES resolutions that aim to protect and conserve species whose survival is threatened by international trade. Since it remains unclear as to what actions would amount to "promote," "encourage," or "endeavor," such hortatory language is also not considered binding or at best, generates low level of legal bindingness for the parties.

In addition, at international level, CPTPP parties are required to cooperate in order to address key issues of environment protection relating to protection of the ozone layer (Article 20.5 of the CPTPP), protection of the marine environment from ship pollution (Article 20.6 of the CPTPP), trade and biodiversity (Article 20.13 of the CPTPP), and transition to a low emission and resilient economy (Article 20.15 of the CPTPP). Put differently, parties shall take concerted international action to tackle important matters including protection of marine species, biodiversity, and the shift to a low-emission economy. For instance, the parties shall cooperate to address matters of mutual interest related to ozone-depleting substances (Article 20.5(3) of the CPTPP), or with respect to pollution of the marine environment from ships (Article 20.6(3) of the CPTPP), or regarding the issue of biodiversity (Article 20.13(6) of the CPTPP). However, since the obligations are in the form of "recognizing the importance" and "cooperation," these provisions generate rather soft obligations with low level of bindingness for parties. They are binding commitments but "that arguably is irrelevant," because there is nothing in Chap. 20 or the CPTPP to specify the manner in which the parties should cooperate, it is hard to "imagine a state being compelled to act cooperatively" (UN Environment, International Institute for Sustainable Development, Environment and Trade Hub, 2016). Even if Article 20.15 stipulates that the parties are obliged to cooperate "as appropriate" to address matters of joint or common interest related to transitioning to a low emissions economy, this requirement may turn out to be "general aspiration" rather than binding commitment or at best, it renders soft obligations to be followed by the parties (Kucik, 2019).

At the other end of the spectrum, binding commitments generate hard obligations with respect to, among others, enforcement of domestic environmental law and policy.

*First*, CPTPP parties are obliged to ensure that high levels of environmental protection are provided for in their laws and policies and especially, the parties are prohibited from *failing to effectively enforce their environmental laws* in a manner affecting their trade or investment relationship. Without even requesting its parties to develop new laws and/or policies, CPTPP seems to simply require its parties to effectively enforce their existing laws on protecting and preserving the environment. Yet, in fact, it does not appear as simple as that. In addition to changing, modifying current laws, adjusting current policies so as to ensure high levels of environmental protection, for Vietnam, the implementation of environmental laws and policies has faced with several constraints due to lacks of appropriate institutions, financial and human resources; insufficient environmental regulations and standards; and especially when general awareness in the society toward environment protection is still very low. In this regard, it should also be noted that there is no specific environmental standard provided for in the CPTPP, the matter is then left at discretion of each party.

*Second*, in the same vein as effective enforcement, CPTPP parties are prohibited to lower their environmental standards, weakening or reducing the levels of protection stipulated in their respective laws and policies, in order to encourage trade or to attract investment, including foreign direct investment (Article 20.3(6) of the CPTPP). More specifically, the parties committed not to waive or otherwise derogate from its environmental laws—which, in such a way, could be detrimental to effective enforcement of their domestic environmental laws and policies—in their trade and investment relationship with other parties of the CPTPP. This requirement again reinforces the obligation to ensure high levels of environment protection with respect to domestic law and policies of each CPTPP countries.

*Third*, in strengthening the fight against subsidies that contribute to overfishing, CPTPP parties are prohibited from granting or maintaining certain types of (i) subsidies for fishing that negatively impact overfished stocks, and (ii) subsidies provided to any fishing vessel for illegal, unreported, and unregulated (IUU) fishing (Article 20.16(5) of the CPTPP). Regarding the first type of fisheries subsidies, while other CPTPP parties shall have to implement this obligation within 3 years from the respective dates of entry into force of the CPTPP for that party, Vietnam can request for an extension of two additional years making a total of 5 years from 14th January 2019, the date which CTPPP entered into force for Vietnam, to implement this obligation thereby bringing any subsidy program into conformity with Article 20.16.5(a) of the CPTPP (Article 20.16(6) of the CPTPP).

*Fourth*, with respect to requirements on domestic transparency, procedural, and institutional aspects, CPTPP parties are obliged to comply with the following obligations: (1) to ensure that persons with a recognized interest under domestic law may request competent authorities to *investigate* alleged violations of environmental laws; having *appropriate access to judicial, quasi-judicial, or administrative proceedings* for the enforcement of its environmental laws which are *fair, equitable, transparent,* and comply with *due process of law*; (2) to provide appropriate sanctions or remedies for violations of the party's environmental laws, including damages, a right to seek injunctive relief or governmental action (Article 20.7 of the CPTPP). At the same time, (3) the parties shall designate and notify a contact point to facilitate communications, coordination among themselves, and especially, (4) each party has to establish and/or maintain a mechanism for the receipt and consideration of written submissions from its persons with respect to the status of implementation of this chapter, and in case the submission asserts that the party is failing to effectively enforce its environmental laws in accordance with CPTPP provisions, the Committee on Environment may exercise its consideration as to whether the matter could benefit from cooperative activities. This obligation effectively requires countries like Vietnam to establish and maintain an appropriate institutional body or mechanism to dealing with public submissions from its individuals, enterprises, or institutions.

*Fifth*, this is not an obligation per se but a possibility that Vietnam may be involved in a dispute in case any party violates its obligations in this chapter. The dispute could then be subject to the dispute settlement procedure of the whole CPTPP as stipulated under Chap. 28. The procedure may start with consultations and when consultations have failed to resolve the matters at hand, establishment of a panel may be requested in accordance with Article 28.7 to provide solutions for the dispute. Until now, Vietnam has never been filed a case against it either at the WTO or within the framework of any FTAs, yet preparation for dispute settlement is always of crucial necessity so that the country is able to preserve its rights and obligations specifically provided for in this chapter on environment.

In a different manner compared to CPTPP, environment-related provisions in EVFTA are covered under the umbrella of sustainable development as stipulated in Chap. 13 of this agreement. EVFTA parties generally affirm their commitments to pursue sustainable development in which environmental protection remains one important element, along with economic and social development. Regarding environment-related obligations, Chap. 13 includes similar commitments as in Chap. 20 of CPTPP relating to: effective implementation of existing MEAs (to which it is a party) (Article 13.5 of the EVFTA); the right to establish their own levels of domestic environmental protection while endeavoring to ensure such high levels through effective enforcement of domestic laws and regulations (Article 13.2 of the EVFTA); prohibitions on certain types of fisheries subsidies (Article 13.9 of the EVFTA); promoting cooperation between parties with respect to trade and environment policies; and other commitments on biodiversity, marine resources, as well as procedural requirements such as to establish a contact point to facilitate communications among parties or transparency, capacity building.

Notably, regarding its scope of application, EVFTA Chap. 13 even includes a provision on climate change, a topic which is not specifically regulated in Chap. 20 of CPTPP. However, without setting any international standards, Article 13.6 of EVFTA simply reaffirms the parties' commitments to effectively implement existing agreements such as the United Nations Framework Convention on Climate Change of 1992 (UNFCCC), the Kyoto Protocol and the Paris Agreement, while promoting the transition to low greenhouse gas emissions and climate-resilient economies with low-emission technology and renewable energy.

Last but not least, EVFTA also differs in the manner which the parties use to resolve their conflicts. While CPTPP parties have access to the agreement's dispute settlement mechanism for any disputes arising under its Chap. 20 on environment, Chap. 15 on dispute settlement of EVFTA does not apply to its respective Chap. 13 on sustainable development. The parties to EVFTA shall only have recourse to the procedure provided for under Chap. 13, which is solely dedicated for settling dispute arising from any issues covered by the same chapter relating to sustainable development, including environmental and labor issues. The procedure includes government consultations and if consultations have failed to satisfactorily resolve the conflict, a Panel of Experts may be convened to make recommendations for the solution of the matter (Article 13.17 & 13.17 of the EVFTA).

## *4.7 Challenges and Implications from Legal and Institutional Perspectives*

It could be observed that more ambitious environment-related provisions are mainly found in North-North or North-South trade agreements, and much less in agreements signed by developing countries only. In fact, this is also one of the most prominent characteristics of new-generation FTAs, as the majority of such agreements include at least one party who is developed country or even among the leading economies such as the United States (US) or the European Union (EU) (Vu & Pham, 2019). And within this relationship, it usually happens that commitments on sustainable developments are suggested by developed country member. In the case of Vietnam and for the conclusion of CPTPP and EVFTA, it is worth noting that EVFTA basically follows the model of new-generation FTA that the EU has imposed on its trade agreements with different partners, such as Korea or Singapore, which provides for a chapter on sustainable development including environment issues and, at the same time, the United States's stance is apparently reflected in the Chapter on Environment from the then TPP to the current CPTPP. As these commitments are in line with developed countries' standards, the inclusion of environment-related provisions may pose challenges to developing countries, including Vietnam, in the implementation of these provisions, especially with respect to binding environmental obligations.

### 4.7.1 For Vietnam Authorities

Vietnam already participated in several MEAs and thus, implementing international commitments on environment protection is not something new. The Law on Environment protection of Vietnam 2020 includes several provisions promoting the application of international agreements to which Vietnam is a signatory in areas relating to, inter alia, controlling and processing of marine and island environment pollution (Article 11 of the Law on Environment Protection of Vietnam 2020), reduction of greenhouse gas emissions or protection of the ozone layer (Article 91, 92 of the Law on Environment Protection of Vietnam 2020), or adaptation to climate change (Chapter VII of the Law on Environment Protection of Vietnam 2020). Yet, for environment-related obligations that are found in new-generation FTAs, the depth and breadth of commitments are at higher levels whereby trade-related drivers, or the interaction between trade and issues of sustainable development, make it more challenging to implement these commitments. Moreover, non-implementation, failure to comply or violations of the obligations could be subject to a dispute settlement mechanism which may even require compensation by the violating party. In such a way, compliance to FTA commitments is of crucial importance for Vietnam. As a matter of fact, however, the country has faced with numerous challenges in this implementation process.

*First*, in establishing and maintaining high levels of protection while striving for effective enforcement of environment protection as mandated for under, for instance, CPTPP, challenges come from low level of awareness when it comes to environmental issues, a serious lack of funds, human resources, and especially a lack of domestic legislative and enforcement capacity for environmental laws and policies (OECD, 2015).

In fact, Vietnamese laws and regulations have basically established the general framework for environment protection in several aspects of the society. However, environment-related issues are regulated in many different laws such as Law on Investment, Law on Construction, Law on Water Resources, Law on Irrigation, etc. and thus, cooperation and enforcement remain difficult, let alone some overlapping and conflicting provisions between these laws. As noted, environmental laws and regulations in Vietnam could point toward the model of Community-Driven

Regulation whereby community actions can and does drive environmental regulation (O'Rourke, 2002). The limit of this model is that community members often end up focusing on "purely localized, short-term, acute impacts of pollution" and tend to push for pollution control rather than prevention due to their lack of capacity (O'Rourke, 2002). In such a way, environmental procedures such as monitoring, impact assessments, and research on environmental issues remain very important for a comprehensive and long-term planning to protect the environment.

With respect to domestic transparency, procedural and institutional aspects relating to environment protection, challenges come from currently weak institutional framework with overlapping procedures.

In addition, there is a serious lack of high-level environment-related technical regulations and standards for products. It is worth noting that new-generation FTAs, including CPTPP and EVFTA, will most likely open markets for trade and investment activities for Vietnam thereby promoting development in areas such as textiles and clothing, electronics, machinery, footwear products, wooden products, seafood products, steel, crude oil, pepper. And without proper environmental regulations and standards, these developments may cause severe adverse impacts toward the environment. Since Vietnam has to opens its market for foreign products, it remains crucial to establish high level environmental regulations up to international standards to prevent imports of products which are harmful to the environment.

Another issue is the lack of enforcement capacity, reflected by the increasing number and the seriousness of environmental violations, including the problems of increasing air, noise, and water pollution, deforestation, and land degradation as well as the growing amount of highly polluting waste (Ortmann, 2017). Put differently, existing laws and regulations of Vietnam on environment protection have not been effectively implemented. Apart from the lack of domestic legislation, reasons also include shortages on human resources, lacking competent staff/officials to carry out duties in relation to trade-related environmental obligations, and a serious lack of funds to introduce and apply environment-friendly technology within the society. Furthermore, a generally low level of public awareness on environmental issues is also one main reason for the increase in environmental violations and makes it more difficult to enforce relevant laws and policies.

*Second*, the problem of "implementation deficit"—referring to the fact that environment-related FTA provisions may look good on paper, but its impacts can be difficult to identify—remains a crucial challenge (OECD, 2015). For developing countries like Vietnam, comprehensive environmental impact assessment is even harder to be conducted due to difficulties in establishing robust baselines for comparisons (APEC Committee on Trade and Investment, 2017). In this regard, it is suggested that Vietnam should take advantages of provisions on international cooperation with other FTA partners to improve the quality of its assessment and also for capacity building purpose.

*Third*, there is a risk that failure to comply with environment-related obligations under new generation FTAs may lead to a dispute to be filed against Vietnam and thus, it remains a challenge that the country has to prepare for through, for instance, enhancing capacity of the government and its environmental agencies as well as improving the institutional framework.

### 4.7.2 For Enterprises

Trade-driven environmental commitments, admittedly, make enterprises become more responsible toward the environment thereby adjust their activities toward sustainability. In this regard, one of the most important challenges for relevant enterprises is the compliance cost to environment regulations. New-generation FTAs, especially CPTPP, provide for ambitious commitments relating to environment protection which would gradually restrict the use of old and unfriendly technology that could be harmful to the environment, and promote the development of new technology using renewable energy or eco-friendly materials for production. In such a way, enterprises have to improve their awareness of environmental provisions in existing FTAs and at the same time, make up for increased compliance cost to invest in new technology, increase the cost for (raw) materials, etc. Compliance cost may lead to increase in prices of final products which could then make the products less attractive to customers, especially in short-term plan.

For export-oriented enterprises, it remains crucial that they have to improve their awareness of foreign regulations on environment protection and comply with more demanding laws and regulations from importing countries. Regardless of industries, enterprises exporting to other CPTPP member countries will possibly be affected by changes in environmental policies in response to commitments under Chap. 20 of this agreement. This is particularly relevant for exports of Vietnamese goods such as agriculture, fisheries, and wood-processing products due to their environmental effects.

Yet, however, environment-related commitments may also open up opportunities for enterprises in transferring environmental technology, manufacturing and exporting environmental goods and services such as environmentally friendly and organic products, or waste and wastewater management services (International Trade Center, 2014). Since trade in environmental goods and services is considered a means to improving environmental and economic performance while addressing global environmental challenges, CPTPP requires parties to endeavor to address any potential barriers to trade in these goods and services, thereby creating meaningful business opportunities for Vietnamese enterprises in this field (Article 20.18 of the CPTPP).

### 4.7.3 For Consumers and Other Stakeholders in the Society

Compliance to environment-related commitments under new generation FTAs would generate environmental benefits to the society as a whole. As noted, these FTAs can actually become a very powerful tool to influence environmental policy thereby leading to economic growth that is less polluting or countering the "pollution haven" effect by providing a level playing field across member countries of FTAs (OECD, 2015).

Since enterprises have to adjust their activities in accordance with environmental regulations towards sustainability, prices of products in affected industries (fisheries, wood processing, etc.) may increase due to an additional compliance cost which then affecting purchasing power of consumers. In this regard, the important point is to raise awareness of the society as a whole with respect to environment protection in general and specific FTA provisions as well as domestic laws and regulations in particular, so as to promote the use of eco-friendly products, even at higher prices.

Also, the community remains an important driver in the implementation of environmental regulations by directly pressuring firms to reduce pollution, pressuring state environmental agencies to improve their monitoring and enforcement capabilities and raise public and elite awareness of environmental issues and the trade-offs between development and the environment (O'Rourke, 2002). Under Article 5 of the Law on Environment Protection of Vietnam 2014, it is also necessary to "facilitate the involvement of organizations, family households and individuals in the environmental protection activities; inspect and supervise the performance of environmental protection activities in accordance with laws."

## 4.8 Some Recommendations for Vietnam in the Implementation of FTA Environmental Provisions toward Sustainable Development

Environment-related commitments within Vietnam's new-generation FTAs, even if not all of them generate binding obligations, can admittedly bring about both advantages and challenges to the country. The challenges, in return, could serve as opportunities for Vietnam to further revise its environmental laws and policies to maintain high levels of environment protection up to international standards, and exercise effective enforcement in order to promote a less polluting economy. Therefore, it remains crucial to combat challenges thereby taking advantages of implementing environment-related obligations, especially those generated by CPTPP and EVFTA.

As a starting point, it is of crucial importance to revise or modify current laws and policies to provide high levels of environment protection in compliance with international commitments under MEAs to which Vietnam is a signatory. Since new generation FTAs, CPTPP, or EVFTA do not include specific environment-related technical regulations or standards, there is also a need to develop a set of environment regulations up to internationally recognized standards to prevent unqualified products that are harmful to the environment to be imported into Vietnam. In this respect, Vietnam could take advantage of provisions promoting environmental cooperation, sharing of best practices to help develop these environmental standards in a manner moving toward more regulatory harmonization between FTA parties. For instance, Japan cooperated with Vietnam to revise Vietnam's environmental laws in 2018, and this cooperation is further enhanced after CTPPP's entry into force.

In the same line, cooperation with other member countries, especially developed economies, remains a useful means which could help strengthen Vietnam's institutional framework to promote environment policies while maintaining effective enforcement. Within the country, implementation could be improved through better communication between government institutions and environmental agencies to support the enforcement and review of environmental policies and regulations.

Moreover, capacity building remains an important part in the implementation of environmental obligations for developing countries like Vietnam toward effective enforcement of international commitments, domestic laws, and regulations. Capacity building could certainly be strengthened with support from more advanced economies in realizing environmental cooperation. In fact, it is considered a key factor to balancing the concern of developed parties toward the weak environmental management of developing parties and thus, developing countries are often able to take advantage of that (APEC Community on Trade and Investment, 2017).

Another recommendation is to promote and raise public awareness of environmental issues and its associated impacts on different stakeholders and the society as a whole. As required by CPTPP Article 20.17, the parties shall promote public awareness of its environmental laws and policies, including enforcement and compliance procedures, by ensuring that relevant information is available to the public. At the same time, it is crucially important to increase financial resources investment in advanced environmental technologies and to enhance government capacity in dealing with environmental inquiries or even disputes.

Last but not least, in order to foster environment protection thereby further promoting the aims of sustainable development, not only enterprises but the whole community should also join forces. The authorities could then develop appropriate policies to encourage enterprises to take initiatives relating to sustainable development, while the community remains proactive in identifying, pressuring individuals, enterprises, and state on environmental violations and thus, forcing environmental agencies to improve their operation and enforcement capacities.

## References

### *Legal Documents*

CPTPP and EVFTA legal texts

Laws on Environment Protection 2005 and 2014 and 2020

National Plan on Environment and Sustainable Development in the 1991-2000 period

Order No. 36/1998/CT-TW dated 25/6/1998 by the Political Bureau on the enhancement of environment protection activities during the national industrialization and modernization period

Sustainable Development Orientation of Vietnam

The Sustainable Development Strategy of Vietnam in the 2011-2020 period

## References

APEC Committee on Trade and Investment (2017) Study report on environmental provisions in APEC member economies' FTAs/RTAs. Available at: https://www.apec.org/Publications/2018/05/Study-Report-on-Environmental-Provisions-in-APEC-Member-Economies-FTAsRTAs

Bodansky D, Lawrence JC (2009) Trade and environment. In: Bethlehem D et al (eds) The Oxford handbook of international trade law. Oxford University Press

Charnovitz S (2013) Trade and the environment. In: Lukauskas A, Stern RM, Zanini G (eds) Handbook of trade policy for development. OUP, Oxford

Fickling M, Hufbauer G (2012) Trade and the environment. In: Daunton M et al (eds) The Oxford handbook on the world trade organization. Oxford University Press

Hofmann C, Osnago A, Ruta M (2017) Horizontal depth: a new database on the content of preferential trade agreements. World Bank Policy Research Working Paper 7981

International Trade Center (2014) Trade in environmental goods and services: opportunities and challenges. Available at: http://www.intracen.org/uploadedFiles/intracenorg/Content/Publications/AssetPDF/EGS%20Ecosystems%20Brief%20040914%20-%20low%20res.pdf

Kucik J (2019) Trade deals have glaring omission: Environmental standards. Available at: https://thehill.com/opinion/energy-environment/422458-trade-deals-have-glaring-omission-environmental-standards

Ngan V, Son N (2019) General theoretical issues of new generation free trade agreements: when the purpose is not just about trade liberalization. Vietnam Leg Sci J 03(124):1–5

O'Rourke D (2002) Community-driven regulation: towards an improved model of environmental regulation in Vietnam. In: Evans PB (ed) Livable cities: urban struggles for livelihood and sustainability

OECD (2001) The development dimensions of trade

OECD (2015) Report on the Regional workshop on 'Regional trade agreements and the environment. http://www.oecd.org/officialdocuments/publicdisplaydocumentpdf/?cote=COM/TAD/ENV/JWPTE(2014)48/FINAL&docLanguage=En

Ortmann S (2017) Environmental governance in Vietnam. Springer Link, pp 10–20

UN Environment (2016) A sustainability Toolkit for Negotiators. Available at: https://www.iisd.org/toolkits/sustainability-toolkit-for-trade-negotiators/3-environmental-provisions/3-1-binding-or-non-binding-commitments/

van Aaken A, Lehmann TA (2011) Sustainable development and international investment law: an harmonious view from economics. In: Prospects in international investment law and policy, vol 2013. Cambridge University Press, Cambridge, pp 317–339

**Ngan-Kim Vu** is currently a WTO Co-ChairHolder and lecturer in law at the Faculty of Law, Foreign Trade University, Vietnam. She graduated with BBA, specializing in International Business Law, from Foreign Trade University, Vietnam; obtained Master of International Law and Economics from World Trade Institute, University of Bern, Switzerland, and Ph.D in Law and Economics from Bocconi University, Milan, Italy. Her areas of expertise include international trade law, WTO-related legal issues, legal aspects of globalization, sustainable development and regional integration. In addition to teaching and research experiences at FTU and as visiting lecturer at several other institutions, she worked as a consultant at the World Trade Organization in Geneva, Switzerland (2011–2013), and has provided consultancy services for several institutions, including NGOs, as well as Vietnamese and foreign enterprises.

**Hang-Thi Minh Hoang** is a Doctor of Law, Faculty of Law, Foreign Trade University, Vietnam. She acquired advanced degrees in international law and economics at the World Trade Institute, University of Bern in 2012 and Bocconi University in 2019. She teaches courses on Introduction to Law, Public International Law, and Research Methods at Foreign Trade University. Her research interests cover trade-related issues in free trade agreements.

# Environment Protection in International Investment Agreements of Vietnam: Investment for Sustainable Development?

**Kim Anh Dao**

**Abstract** Foreign investment has great contribution to the economic development of the host countries; however, accompanied with the flow of capital are the risks of environment pollution and resource degradation. While international investment agreements (IIAs) aim to protect economic interests by promoting and encouraging investment, the question arises as to the role of noncommercial values such as environment in IIAs. The article examines the possible conflicts between measures to protect the environment implemented by host States with their obligations of investment protection under IIAs and how such conflicts are resolved under IIAs concerning Vietnam. Through the survey of 66 bilateral investment treaties and several treaties with investment provisions which Vietnam has negotiated and/or participated in since 1990s, the study reveals the emerging trend of integrating environmental concerns into investment treaty-making practices, using a variety of environment languages in the preamble as well as in the text of IIAs. Accordingly, it provides some thoughts and recommendations for Vietnam in drafting and negotiating IIAs toward a more sustainable investment policy.

## 1 Introduction

Investment has been long perceived to be of significant importance for sustainable development. In particular, the flow of foreign investment can provide capital, technology, and skilled labor to develop the economy, upgrade infrastructure, and promote social equality which are key requirements for promotion of sustainability (Bernasconi-Osterwalder et al., 2012, p. 1). Therefore, countries worldwide have actively participated in negotiating and concluding international investment treaties

K. Dao (✉)
Foreign Trade University, Ha Noi, Vietnam
e-mail: anhdk@ftu.edu.vn



H.-T. Vu, L.-P. Dana (eds.), *Vietnam on the Road to Sustainable Economic, Social and Business Development*, Frontiers in South and Southeast Asian Development Research, https://doi.org/10.1007/978-981-96-4371-4_7

with the promotion and protection of foreign investors becoming one of their major concerns.

However, facts have proved that not all investments can serve the goal of sustainable development. Unfortunately, the dramatic increase of foreign investment has resulted in certain threats to the host States, particularly causing serious environment pollution and degradation of natural resources. These problems have alerted nations to implement stricter measures with an aim to protect their natural environment against the pursuit of investors' interests. These measures in some cases may cause conflicts with the standard of treatment provided for investors under investment treaties, and even worse lead to legal disputes between foreign investors and the host States. As a result, the question of how to resolve the conflicts between environmental values and economic benefits of attracting investment has posed a great challenge to countries in drafting and negotiating international investment agreements (IIAs). The approach to integrate environmental issues into the world of international investment law may vary from country to country, with those which opened up recently to foreign investment, still searching for the right balance.

The latter has been the case in Vietnam. Since the shift from a centrally planned to a market economy in the last few decades, especially after becoming a member of the World Trade Organization (WTO) in 2007, Vietnam has made tremendous efforts to attract foreign investment through participating and signing investment treaties at both bilateral and multilateral levels. Nevertheless, the recent environment scandals in foreign investment projects such as the case of Vedan company discharging untreated waste into the Thi Vai River or Formosa Ltd. causing marine life disaster in the central region of Vietnam have urged the government to impose higher control over the environmental impact of foreign investment. As a result, it is vital for Vietnam to adopt environmental measures in consistency with its obligations committed under IIAs. With the increasing demand of Vietnam to attract high-quality investment without the trade-off between economic benefits and the risk of environmental pollution, there has been much higher pressure from scholars, legal practitioners and the business sector for reconsidering environment-related provisions in IIAs of Vietnam and assessing their implications on our domestic law-making process.

While the impact of foreign investment on the environment has received attention for decades, literature on the environment provisions under IIAs has proliferated in recent years. A number of researchers have highlighted the increasing need for integrating public values including the environment into the world of investment treaties. Gordon and Pohl (2011) reveals a statistical picture over governments' investment treaty writing practices in relation to environmental concerns in a sample of 1623 IIAs which made up around 50% of the total number of IIAs. The survey shows that although only about 8% of the chosen treaties mentioned environmental concerns, such language has increasingly appeared in investment treaties over time with wide variations across countries and time (Gordon and Pohl, 2011, p. 3). Meanwhile, from the perspective of developing countries, (Tienhaara, 2006), based on practical analysis of several controversial investor-state disputes concerning environment measures, points out the dilemma in which the protection provided to investors may restrict the governments' ability to regulate investment for the environment protection, leading to a so-called chilling effect on government policy. In addition, Benedetto (2013) offers an extremely detailed examination of the problematic impact of environment issues on international investment law from both perspectives of treaty-making and arbitral dispute settlement. It also demonstrates the heated debate over fragmentation between international environment law and investment law and examines possible methods for reconciling investor benefits and environment concerns through the proposal of the exception models. In a search for the right balance, Camille (2017) suggests that the notion of sustainable development can offer a solution to better articulate the interplay of international investment law and environment law, producing a concrete shift in investment policies. With a detailed analysis of environmental protection incorporated in newly drafted IIAs, the article argues that these provisions are expected to provide more coherency and legitimacy to integrate environment concerns directly into the text of treaties, therefore hopefully providing countries a defense against investor's claims before arbitration.

In spite of the heated debate over the inclusion of environmental values in investment treaties worldwide, there has been little literature on this issue with regard to the case of Vietnam. Nguyen (2018b) adopts a case study method to examine a number of environmental disputes between businesses and citizens including those involving foreign investment projects and highlight gaps in domestic regulations to resolve these disputes. However, it has yet to put these problems in the context of international law which requires domestic law to be consistent with international standards. Nguyen (2018a), despite exploring difficulties facing the government in regulating environment issues related to foreign investment, places more focus on expropriate provisions under IIAs.

Given this context, this research is expected to fill the existing gap in Vietnam by offering theoretical knowledge about the fundamental nature and classification of environment provisions under IIAs. The research seeks to discuss and evaluate different options for Vietnam to keep a balance between promotion of foreign investment and protection of the environment during the process of negotiation and conclusions of IIA. With this objective, the main research question concerns potential conflicts between environment values and investors' interests protected under IIAs concerning Vietnam. With a thorough comparative analysis of more than 70 IIAs involving Vietnam, it will formulate detailed proposals in terms of both investment treaty negotiation and implementation in order to improve the quality of investment for environment protection, and further for the ultimate goal of sustainable development.

The research begins with a general analysis of the challenges to regulate environment issues and their interaction with traditional standards of treatments in IIAs. Further, it will evaluate the recent development of environment-related provisions in investment law and their application in dispute settlement by arbitration. In the next parts, it will analyze Vietnam's approach in dealing with environment protection in IIAs, compare and contrast the approaches taken under different treaties. The analysis also puts the Vietnamese approach in comparison with recent reforms of investment treaties worldwide with an aim to reveal an outlook for international harmonization. Finally, the chapter will give a suggested approach for Vietnam in dealing with this matter in the both negotiation and implementation of investment treaties.

## 2 Environment Values in the Context of International Investment Law: Potential Conflicts with Traditional Standards of Treatment

Although there might be differences in languages, in general, IIAs are designed to ensure the fundamental principles regarding the treatment for investments and investors of contracting parties. These principles include (1) nondiscrimination, (2) fair and adequate treatment, and (3) protecting investors from unlawful expropriation. However, facts have shown that environmental measures carried out by host States in some cases may raise conflicts with or even violate the abovementioned principles. As a consequence, claims by the affected investor are likely to be raised against such environmental measures in disputes with the State (Slater, 2015, p. 146). The common situation that often occurs in international investment disputes is when the State changes or issues new policies and regulations to lift environmental standards or encourage the thrive of environmentally friendly industry sectors. These changes may adversely affect investments (e.g., increasing costs and causing discrimination among investments). In addition, measures (such as administrative sanctions, revocation of investment permits) to deal with violations of environmental laws by foreign investment projects may also be sued by investors for breaching standards of treatment in IIAs. This chapter will analyze some investor-State disputes which show potential conflicts between environmental protection issues and traditional principles of treatment in IIAs.

### *2.1 S.D. Myers, Inc. v. Government of Canada*

The first case concerns the principle of nondiscrimination in the North American Free Trade Agreement (NAFTA). In this case, the claimant, S.D. Myers Group (based in the United States), is the leading corporation in the field of collection and disposal of PCB waste[1] in the United States. In the 1990s, S.D. Myers decided to invest in Canada, and its main business activities were to set up facilities in Canada

[1] Polychlorinated biphenyl (PCBs) is a group of artificial compounds which used to be widely used in electrical equipment. PCBs was banned in the late 1970s in numerous countries due to the risks of causing the environment and health harm. However, PCBs is a very stable compound, thus the treatment requires modern methods and technologies, which leads to the appearance PCBs collection and disposal industry.

to collect PCB waste, then transferring them to the factory in the United States for disposal. In 1995, Canada's Minister of Environment issued a decision banning the export of waste containing PCBs, causing S.D. Myers' project to stop operating. The claimant argued that Canada's measure had created more favorable treatment for domestic companies and aimed to protect them from S.D. Myers' market dominance, since the waste disposal cost offered by S.D. Myers in the United States had been much lower than that offered by local companies in Canada. Canada argued that the PCBs export ban has been enforced equally without discrimination among any investors, and aimed to limit the risk of environmental harm during the cross-border transportation of PCBs. This measure also aims to comply with Canada's commitment under the 1989 Basel Convention on the Control of Transboundary Movements of Hazardous Wastes and Their Disposal.

In this particular case, the arbitral tribunal upheld that Article 1102 of NAFTA on national treatment should be interpreted in harmony with NAFTA's general principles, which emphasizes the "states have the right to establish high levels of environmental protection" and "environmental protection and economic development can and should be mutually supportive" (S.D Myers, Inc. v. Government of Canada, 2002, p. 247). Despite the foregoing principle of interpretation, when considering the merits of the dispute, the arbitral tribunal paid little attention to Canada's argument for protection of public interests and instead concluded that Canada's measures have breached Article 1102 of NAFTA. The arbitral tribunal merely relied on economic factors to conclude that S.D. Myers and other Canadian waste disposal companies have been in like-circumstance and thus the ban on PCBs export has indeed created benefits solely for local companies (S.D Myers, Inc. v. Government of Canada, 2002, p. 251). Meanwhile, should the environmental factors be considered, the cross-border transport of waste carried out by S.D. Myers would be more likely to cause harm to the environment and human health than local waste collection and disposal activities.

In addition, the arbitral tribunal also commented that Canada could have achieved its policy goals by numerous measures; however, imposing administrative decisions to ban PCBs export is not a measure in conformity with investment regulations under NAFTA (S.D. Myers, Inc. v. Government of Canada, 2002, pp. 255–257). Therefore, the obligation to ensure non-discrimination against foreign investors could become an obstacle for the State in enacting and enforcing environmental protection measures. This conflict is expected to escalate in the future as countries are making efforts to enforce public policies, such as promoting renewable energy, tightening regulations on food and biosafety safety study, etc. These policies would create advantages for particular entities, such as green energy corporations, thus it is likely that States may encounter opposition from investors in traditional industries, who demand nondiscrimination treatment among investors (Camille, 2017, pp. 532–535).

## 2.2 *Técnicas Medioambientales Tecmed, S.A. v. The United Mexican States*

In addition to the principle of nondiscrimination, the fair and equitable treatment (FET) is also a widely accepted standard of protection in the majority of IIAs. In Técnicas Medioambientales Tecmed, S.A. v. The United Mexican States (*Tecmed* case), the investor has brought claims against Mexico's enforcement of environmental regulations due to violation of FET. In this case, the Spanish investor, Tecmed S.A, via a public auction, purchased Cytrar—a landfill of hazardous industrial waste from the authority of Hermosillo province (Mexico). At the time of auction, Cytrar had a 5-year permit and could be renewed from time to time. Subsequently, due to local residents' opposition because of environmental pollution, the Mexican government negotiated with the investor to relocate the waste disposal landfill to a different location with a "promise" that Cytrar would remain in operation until the new location is ready. However, Cytrar's permit extension was then rejected. The arbitral tribunal opined that Mexico had breached FET and, in particular, had frustrated the investor's legitimate expectations due to the following reasons:

> [FET clause] requires the Contracting Parties to provide international investments treatment that does not affect the basic expectations that were taken into account by the foreign investor to make the investment. The foreign investor expects the host State to act in a consistent manner, free from ambiguity and totally transparently in its relations with the foreign investor, so that it may know beforehand any and all rules and regulations that will govern its investments, as well as the goals of the relevant policies and administrative practices or directives, to be able to plan its investment and comply with such regulations. Any and all State actions conforming to such criteria should relate not only to the guidelines, directives or requirements issued, or the resolutions approved thereunder, but also to the goals underlying such regulations. The foreign investor also expects the host State to act consistently, i.e. without arbitrarily revoking any preexisting decisions or permits issued by the State that were relied upon by the investor to assume its commitments as well as to plan and launch its commercial and business activities (Técnicas Medioambientales Tecmed, S.A. v. The United Mexican States, 2003).

In the *Tecmed* case, the arbitral tribunal seemed not to give due attention to the environmental goals that the Mexican government's measures sought to achieve. The above dispute shows that the conflict between investors' interests and environmental benefits not only occurred during the stage of issuing policies made by the State, but also could arise directly from the enforcement of policies conducted by local government agencies.

### 2.3 *Compañia del Desarrollo de Santa Elena S.A. v. Republic of Costa Rica*

This is a case where an environmental measure of the State was alleged to have violated the expropriation provision under the related IIA. The risk of being expropriated of assets is one of the biggest concerns for foreign investors. Therefore, provisions to protect investors from expropriation are an integral part of IIAs. However, it should be noted that IIAs do not absolutely prohibit the State from conducting expropriation, but require the State to meet certain conditions for such measures to be deemed lawful. The first condition is that the measure must serve for public purposes. Since public purposes are not clearly defined in IIAs, this concept is interpreted broadly and differently based on the conditions of each country, and often covers the purposes of environmental protection, public health, and national security. Secondly, the measure must be taken neither arbitrarily nor in a discriminatory manner. Thirdly, enforcement of the measure must follow proper legal procedures. Finally, when the State enforces this measure, there must be appropriate compensation for investors.

In Compañia del Desarrollo de Santa Elena S.A. v. Republic of Costa Rica (*Santa Elena case*), the investor initiated claims against Costa Rica for taking over the investor's land to build a wildlife park, in exchange for an inadequate compensation determined by the State. The arbitral tribunal did not deny the fact that the State deprived the land for public purposes, however argued that this purpose did not change the expropriation nature of the measure and does not exempt the State from the obligation to compensate for the investor. According to the arbitral tribunal, "expropriatory environmental measures—no matter how laudable and beneficial to society as a whole—[with respect to international investment laws] are similar to any other expropriatory measures" (Compañia del Desarrollo de Santa Elena S.A. v. Republic of Costa Rica, 2000, p. 192). Under this approach, the arbitral tribunal tends to separate environmental issues from the principles of treatment in IIAs.

Meanwhile, IIAs protect investors from not only direct expropriation, but also indirect expropriation, that is, the measures that do not necessarily lead to any transfer of title, but result in an effect equivalent to the expropriation of investor's ownership (Trinh, 2017, p. 138). The environmental measures of a host State, whether general measures such as the ban on harmful products, or specific measures such as revocation of permits, may materially affect the foreign investor's ownership title (Camille, 2017, p. 541). As a result, such environmental measures, which are put under high risk of being sued by the investor, may be deemed indirect means of expropriation. In the *Tecmed* case, for example, the rejection of permit extension for the landfill has removed all economic and commercial activities of the investment, thus is deemed to constitute an indirect expropriation of investors. These disputes raise concerns among countries that the protection provided to investors in IIAs may be abused to put a burden on the States when they enforce their public policies.

Thus, when countries have been paying more attention to environmental measures, it also increases the risk of infringing foreign investors' interests which are protected under IIAs. This is part of the conflict between the protection of foreign investors and the State's right to regulate, and in the broad sense, the conflict between the public interests and the private interests. Consequently, when facing the risk of being sued and having to compensate investors, countries might experience a "chilling effect" in enacting and changing laws, despite that new laws would benefit their society (R. Gray, 2002, p. 311). Then, the issue in question is whether IIAs can anticipate these conflicts and, if so, how such issues have been resolved under the current system of investment treaties.

# 3 Development of Environment Provisions in International Investment Law

## 3.1 The History

While noncommercial issues such as environmental concerns have long been mentioned in free trade agreements, this problem seems to attract little attention in negotiation for IIAs, especially those concluded before 2000. According to statistics of UNCTAD, despite the gradual increase over time, by 2000 the rate of IIAs which mentioned environmental concerns only accounted for around 8%—a relatively small proportion of the total number of concluded IIAs (see Table 1). Another survey conducted by the OECD also showed similar results. This study revealed that environmental issues which first appeared in the BIT between China and Singapore in 1985 remained relatively rare in investment treaties until the mid-1990s. The number of IIAs that mention environmental issues increased gradually from 2002, and by 2008, nearly 90% of the newly signed agreements have included such languages (Gordon and Pohl 2011, p. 5).

The absence of express environment provisions in old-generation IIAs can be explained by a number of reasons. Firstly, in terms of history, the appearance and purpose of IIAs stemmed from the demands of (1) on one hand, capital-exporting

**Table 1** The number of IIAs with environment languages in the period of 1985–2005

| Period | Before 1985 | 1985–1990 | 1991–1995 | 1996–2000 | 2001–2005 |
|---|---|---|---|---|---|
| The number of IIAs with reference to environmental aspects in preamble | 0 | 0 | 7 | 10 | 33 |
| The number of IIA with environmental language (any mentioning in the text, except preamble) | 16 | 7 | 17 | 53 | 67 |
| Total number of new IIAs concluded in the period | 222 | 179 | 597 | 732 | 469 |

Source: author's compilation based on UCNTAD's IIA Mapping Project (UNCTAD, n.d.-b)

countries that wished to protect their investors when carrying out investments abroad, and (2) on the other hand, capital-importing countries that sought to attract foreign investment to motivate the development of their domestic economics. Therefore, with their central concerns over encouraging and protecting investment, countries when negotiating IIA seemed to give more focus on commercial values and economic benefits. Consequently, noncommercial values such as environmental concerns are of little interest (Camille 2017, p. 529). Second, it was initially believed that in general, foreign direct investment (FDI) would enhance the social welfare in host States by means of not only capital flows but also transfer of technology, which would create jobs and improve living standards (Camille, 2017, p. 530). As a result, it seemed that countries had failed to anticipate or deliberately left out the negative sides of FDI.

However, the more investment activities develop, the more negative effects of FDI are revealed. With an increasing amount of investment disputes arising from the State's environmental measures, the lack of environment provisions under IIAs is considered a major gap to the current system. Consequently, investor-State disputes concerning environment issues have led to unexpected and unpredictable outcomes which greatly depend on the arbitral tribunal's viewpoints. Some arbitral tribunals hold that the provisions of IIAs aim at the core objectives of protecting and encouraging investment, thus being completely independent from other fields of law including environment law. In other words, environmental value is claimed to play a minor role in the area of international investment law as it is not the benefits that IIAs seek to protect. The arbitral award in the *Santa Elena* case is a typical example of this point of view. In this case, the respondent argued that their measure was to fulfill the country's international obligations to protect unique ecological sites. The arbitral tribunal, however, rejected this argument, arguing that international regulations on the obligations to protect the environment do not change the nature of a measure that deprived investors of their rights (Compañia del Desarrollo de Santa Elena S.A. v. Republic of Costa Rica, 2000, p. 192). Such a strict approach has separated international investment law from its correlation with other fields of law and poses a great risk for countries in enforcing their international commitments under different areas of laws (Benedetto, 2013, p. 139).

In conclusion, the enforcement of environmental protection measures is generally believed to bring benefits to society. However, it may also have some adverse effects on certain groups, particularly businesses and investors, who are highly exposed to such effects. In terms of investment law, the issue in question is who will incur costs due to those adverse effects, whether it is investors to suffer losses for the public interests or the State must compensate for the losses incurred by investors? This issue needs to be resolved in IIAs to ensure consistency in application and interpretation by arbitrators, which then increases the predictability of international investment law. The challenge is how to include public values such as the environment into IIAs where private interests of investors should be respected and protected.

## 3.2 Trend Forward

The appearance of noncommercial values is considered as a new trend in negotiating and drafting IIAs (Camille, 2017, p. 536). The UNCTAD study shows that the number of IIAs with environmental provisions has risen over time (UNCTAD, n.d-b). While in old-generation IIAs (prior to 1999), the environment issue seemed to be put separately from the treaty provisions (merely got mentioned in the preamble of about 1% of IIAs), the appearance of such issue increased rapidly in the period from 2000 onward. Further, not only emphasizing the role of environmental protection in the preamble, these IIAs also integrated this matter in specific clauses. According to UNCTAD's statistics, while only 92 out of 1708 IIAs concluded in the period prior to 2000 had express provisions on environment concerns (around 5%), in the period from 2000 to 2019, the proportion went up to 24%.

In addition, the OECD study also illustrates that the majority of IIAs that include environmental provisions are free trade agreements with investment chapters while BITs, especially old-generation BITs (concluded before 2000) usually lack this language (Gordon and Pohl, 2011, p. 5). This can be explained by the fact that noncommercial values such as environment and public health have long been taken into consideration during the negotiation of free trade agreements, therefore has a pervasive influence on international investment law.

In terms of content, environmental clauses in investment treaties are diverse and designed according to the investment attraction policy pursued by the contracting States. Based on the policy purpose, the OECD survey divides environmental languages in IIAs into the following seven categories (Gordon and Pohl, 2011, p. 6):

1. General language in preambles

   This is one of the most frequently observed categories of environmental language identified by the OECD in its survey. The language used may be diversified which reflects a wide range of concerns including references to the concept of sustainable development or affirmations that the promotion of foreign investments should be achieved without relaxing environmental standards or recognition of the increasing need for measures to protect the environment, etc.

2. Reserving policy space for environment regulation

   Traditionally, a number of investment treaties include provisions to retain the State's policy space to regulate environmental issues. The North American Free Trade Agreement, for example, in Article 1114 (1) states that:

> Nothing in this Chapter shall be construed to prevent a Contracting Party from adopting, maintaining or enforcing any measure otherwise consistent with this Agreement that it considers appropriate to ensure that investment activity in its territory is undertaken in a manner sensitive to environmental concerns.

Besides, another recently developed type of language, often referred to as "general exceptions" clause, is borrowed from the concept of general exceptions in international trade law which expressly put environment concerns as a general exception to obligations under the treaties. These exceptions which are often modeled on the

Article XX of the General Agreement on Tariffs and Trade (GATT) allow contracting Parties to retain some leeway in the implementation of general regulatory measures, including environment regulations that would otherwise be inconsistent with the State's obligations under the treaties. These provisions are increasingly used in the investment chapter of FTA, whereas their appearance in BITs is still limited (Camille, 2017, p. 537).

3. Reserving policy space for environmental regulation for more specific, limited subject matters

   Instead of reserving the policy space as a general principle, a small number of treaties specify such reservation only for specific and limited areas. Therefore, the State's right to regulate environmental matters does not stand independently, but is included as an exception to a provision in the treaty. An example is the Canada–Peru BIT (2006) which provides an exception of performance requirement provision:

> A measure that requires an investment to use a technology to meet generally applicable health, safety or environmental requirements shall not be construed to be inconsistent with paragraph 1(f). For greater certainty, Articles 3 and 4 apply to the measure (Article 7, paragraph 2).

4. Indirect expropriation

This group of treaty precludes nondiscriminatory regulation on environment issues as a basis for claims of indirect expropriation. Some countries include such language in the expropriation clause to avoid the risk that an environmental measure can be exposed to compensation claims of investors for "indirect expropriation".[2]

5. Not lowering environmental standards

   In some treaties, the parties use such language to discourage relaxation of environmental standards to attract foreign investment. The policy purpose of this clause is "to promote the continuous enhancement of the domestic environmental framework of the parties, or at least, to prevent a race to the bottom" between countries (Camille, 2017, p. 543).

6. Environmental matters and investor-State dispute settlement

   While some BITs specify different procedures to deal with an investor-State dispute concerning environment issues, a very limited number of treaties even exclude disputes regarding environment concerns from the application of investor-State dispute mechanism.

7. General promotion of progress in environmental protection and cooperation.

[2] See the US Model BIT 2004, Annex B which states that:

> The Parties confirm their shared understanding that: [...] Except in rare circumstances, nondiscriminatory regulatory actions by a Party that are designed and applied to protect legitimate public welfare objectives, such as public health, safety, and **the environment**, do not constitute indirect expropriations.

A number of clauses that provide a general call for the enhancement of environment standards appear in BITs. This type of clause, however, is often mentioned without a specific link to other treaties' provisions and clauses.

The common language of this group can be found in Article 7, paragraph 1 of the Belgium/Luxembourg–Colombia BIT (2009) which states that "Each Contracting Party shall strive to ensure that its legislation provides for high levels of environmental protection and shall strive to continue to improve this legislation."

# 4 Environment Protection Under International Investment Agreements of Vietnam

Since the first BIT concluded with Italy in 1990, Vietnam has concluded 67 BITs and 25 treaties with investment provisions (TIPs) (UNCTAD, n.d.-a). The study on 67 BITs and 6 selected TIPs[3] reveals that the number of treaties with the environment language of Vietnam remains limited (only 13 out of 73 selected treaties mention environment concerns). Interestingly, only 7 out of 67 BITs[4] have expressly mentioned environmental value (either in the preamble or in the clauses) and most of them are concluded after 2000. In other words, environmental concerns seemed to be absent in most of the old-generation BITs concluded by Vietnam in the 1990s. By contrast, the group of treaties concluded after 2000, especially TIPs, sees more frequent appearance of environment languages. The language is also varied and can be categorized in:

As can be seen from Table 2, there are six types of environment languages in 67 BITs and six TIPs concluded by Vietnam. While a number of treaties put environment concerns under the preambles, others have environment matters connecting with (1) clauses to reserve policy space for environment regulation (either for general application or for specific, limited subject matters), (2) not lower environment standards, (3) investor-State dispute settlement, and (4) indirect expropriation. The following section will analyze each of these types in more detail.

[3] The selected TIPs include the EU-Vietnam Investment Protection Agreement 2019 (EVIPA), Comprehensive and Progressive Agreement for Trans-Pacific Partnership 2018 (CPTPP), Korea–Vietnam Free Trade Agreement 2015, Agreement Establishing the ASEAN-Australia-New Zealand Free Trade Area 2009 (AANZFTA), ASEAN Comprehensive Investment Agreement 2009, and US–Vietnam Trade Relations Agreement 2000.

[4] They include the Estonia–Viet Nam BIT 2009, Venezuela, Bolivarian Republic of–Viet Nam BIT 2008, Mozambique–Viet Nam BIT 2007, Japan–Viet Nam BIT 2003, Turkey–Viet Nam BIT 2014, Singapore–Vietnam BIT 1992, Turkey–Vietnam BIT 1991, and Morocco–Vietnam BIT 2012.

**Table 2** Types of environment-related provisions under IIAs of Vietnam

| No. | Types of environment provisions | BITs | TIPs |
|---|---|---|---|
| 1 | General language in preambles | Estonia–Viet Nam BIT (2009)<br>Venezuela, Bolivarian Republic–Viet Nam BIT (2008)<br>Mozambique–Viet Nam BIT (2007)<br>Japan–Viet Nam BIT (2003)<br>Turkey–Viet Nam BIT (2014)<br>Finland–Vietnam BIT (2008) | EVIPA |
| 2 | Reserving policy space for environment regulation | Japan–Vietnam BIT (2003)<br>Singapore–Vietnam BIT (1992)<br>Turkey–Vietnam BIT (2014) | CPTPP<br>EVIPA<br>AANZFTA |
| 3 | Reserving policy space for environmental regulation for more specific, limited subject matters | Morocco–Vietnam BIT (2012) | CPTPP<br>US–Vietnam Trade Relation Agreement (2000) |
| 4 | Indirect expropriation | | CPTPP<br>Vietnam–Korea FTA (2015)<br>AANZFTA |
| 5 | Not lowering environmental standards | Japan–Vietnam BIT (2003) | |
| 6 | Environmental matters and investor-state dispute settlement | | EVIPA<br>ASEAN comprehensive investment treaty |

Source: Compilation by the author

## 4.1 *General Language in Preambles*

This is the dominant type of environment language appearing in IIAs of Vietnam. Through the preamble of IIAs, contracting States often assert to ensure the harmony between economic objectives (such as promoting and protecting investment) and environmental protection goals (Benedetto, 2013, p. 151). For example, in the Estonia–Vietnam BIT (2009), the preamble emphasizes that "these objectives [of the treaty] can be achieved without undermining health, safety and environmental measures of general application." Likewise, the preamble of the Finland–Vietnam

BIT (2008) affirmed that the contracting parties' pursuit of the BIT's goals does not lower environmental standards. In addition to such a general language, a few IIAs refer to international environmental treaties in their preamble, thereby acknowledging the relationship between international investment law and other relevant fields of international law. One typical example derives from the EVIPA where in the preamble, contracting States:

> Determined to strengthen their economic, trade and investment relationship in accordance with the objective of sustainable development, in its economic, social and environmental dimensions, and to promote investment under this Agreement in a manner mindful of high levels of environmental and labour protection and relevant internationally recognised standards and agreements.

This preamble also puts environment concerns in the context of sustainable development and emphasizes the interrelation between investment promotion and environment protection.

While these preambles place environment matters in relation with the treaties' objectives, it is unclear to define a hierarchy between environment protection and other purposes of the treaties which, in most cases, include protection and promotion of investment. Therefore, it can be argued that the mention of environmental values in IIA's preamble, however, is more of a "statement" rather than have a practical value. The preamble itself does not provide obligations of contracting States but rather can be referred to as the "context" for interpretation of ambiguous terms of the treaty.[5] Therefore, it greatly depends on the arbitral tribunal to determine the role of such statement in the preamble in interpreting IIA's clauses. Some arbitral tribunals have a fairly strict approach to noncommercial values, arguing that this should not be the central goal which IIAs seeks for.[6] For example, in the *S.D. Myers* case as analyzed in part 2, although the preamble of the NAFTA confirms that the pursuit of the contracting parties' goals must harmonize with their environmental protection goals, when considering the ban on chemical substance export conducted by Canada, the arbitral tribunal paid little attention to environmental concerns and instead mainly focused on the economic impact of the State's measure on the investment activities.

---

[5] Articles 31.1 and 31.2 of the Vienna Convention on the Law of Treaties states a general rule for interpretation of treaties that:

1. A treaty shall be interpreted in good faith in accordance with the ordinary meaning to be given to the terms of the treaty in their context and in the light of its object and purpose.
2. The context for the purpose of the interpretation of a treaty shall comprise, in addition to the text, including its preamble and annexes: [...].

[6] See Metalclad Corporation v. The United Mexican States, ICSID Case No. ARB(AF)/97/1; SGS Société Générale de Surveillance S.A. v. Islamic Republic of Pakistan, ICSID Case No. ARB/01/13.

# 5 Reserving Policy Space for Environment Regulation

## *5.1 "Right to Regulate" Clause*

A number of IIAs of Vietnam include clauses in the body to reserve the State's policy space to regulate environmental matters. While this type of clause can be found in CPTPP, EVIPA, and the recently signed Turkey–Vietnam BIT (2014), the scope of the environmental concerns that such treaties describe show some variations. The Comprehensive and Progressive Agreement for Trans-Pacific Partnership (CPTPP) (Article 9.16) refers to "environment concerns" in general, stating that:

> Nothing in this Chapter shall be construed to prevent a Party from adopting, maintaining or enforcing any measure otherwise consistent with this Chapter that it considers appropriate to ensure that investment activity in its territory is undertaken in a manner sensitive to environmental, health or other regulatory objectives.

At first glance, it seems that this provision confirms that commitments in the field of international investment do not hinder the State's right to issue and adjust its domestic environment regulations. However, the language referred to in Article 9.16 of the CPTPP is considered to be political statement rather than a legal norm, since this Article merely confirms that countries have the right to enforce environmental measures that are "appropriate" in light of the Treaty (Camille, 2017, p. 540). In other words, this is a mere reminder of a right that states apparently have, without expansion (e.g., allowing a Contracting Party to, under certain circumstances, enact measures for public purposes that may not be consistent with the treaty). As a result, there is great doubt that in practice, the above provisions can provide guidance for the arbitral tribunal to resolve the conflict between the public interest and the investors' interests (Camille, 2017, p. 542).

A more reasonable approach is found in Article 2.2 (1) of EVIPA, whereby "The Parties reaffirm the right to regulate within their territories to achieve legitimate policy objectives, such as the protection of public health, safety, environment or public morals, social or consumer protection or promotion and protection of cultural diversity." To clarify how the State's right to regulate operates in relation with the State's obligations of investment protection, Article 2.2 (2) further provides that:

> This chapter [Chapter 2. Investment protection] shall not be interpreted as a commitment from a Party that it will not change the legal and regulatory framework, including in a manner that may negatively affect the operation of covered investments or the investor's expectations of profits.

While the CPTPP and EVIPA make reference to the environment in general without specifying the scope, some BITs of Vietnam elaborate on the scope of the reservation of policy space with the list of "beneficiaries" including human, animal, and plant life. For instance, the Singapore–Vietnam BIT (1992), the only old-generation BIT in the 1990s having this kind of provision, affirms the State's right to apply prohibitions or restrictions or other actions directed to "the prevention of diseases and pets in animals and plants," In comparison, the Turkey-Vietnam BIT (2014), while also mentioning specific environmental concerns such as 'the protection of human, animal or plant life" and "the conservation of living or non-living exhaustible natural resources," requires the State's environment measures to be nondiscriminatory.[7]

## 5.2 *"General Exceptions" Clause Modeled on the Text in WTO Agreements*

Given the challenge of maintaining a balance between nontrade values and commercial interests in international investment law, a common trend in drafting the new-generation IIA is to include general exceptions, which allows states to enforce measures that might not be in compliance with IIA regulations to pursue a number of public purposes, including environmental protection (Dordi and Nguyen, 2017, p. 226). General exceptions in IIAs are greatly influenced by general exceptions in free trade agreements, especially from Article XX of the General Agreement on Tariffs and Trade and Article XIV of the General Agreement on Trade in Services (GATS) in the WTO. A number of IIAs that Vietnam has signed recently follow this trend. For example, Article 17. General exceptions of the ASEAN Comprehensive Investment Agreement (2009) adopted almost all of the languages in Article XX of the GATT that:

> 1. Subject to the requirement that such measures are not applied in a manner which would constitute a means of arbitrary or unjustifiable discrimination between Member States or their investors where like conditions prevail, or a disguised restriction on investors of any other Member State and their investments, nothing in this Agreement shall be construed to prevent the adoption or enforcement by any Member State of measures: [...]
>
> (b) necessary to protect human, animal or plant life or health;
>
> [....]
>
> (f) relating to the conservation of exhaustible natural resources if such measures are made effective in conjunction with restrictions on domestic production or consumption.

In addition, instead of adopting similar contents in the general exception of the GATT or the GATS, some IIAs refer directly to these terms. An example is Article 1, Chapter 15 of the ASEAN-Australia-New Zealand Free Trade Area Agreement refers to Article XIV (general exception) in the GATS, which constitutes part of the provisions of Chapter 11 (Investment).

---

[7] Article 4.1 of the Vietnam–Turkey BIT 2014 provides that:

1. Nothing in this Agreement shall be construed to prevent a Contracting Party from adopting, maintaining, or enforcing any non-discriminatory measures:

(a) Designed and applied for the protection of human, animal, or plant life or health, or the environment.
(b) Related to the conservation of living or nonliving exhaustible natural resources.
(c) Imposed for the protection of national treasures of artistic, historic, archeological value.

As modeled on the GATT or GATS, the general exception clause is usually divided into two parts. The first part sets out some conditions on how to apply exceptional measures (in a nondiscrimination, not arbitrarily, unreasonably manner, or without disguised barriers to investors) to prevent the possibility of the clauses being abused by contracting Parties. The second part lists the scenarios and policy purposes that exceptions can be applied to (Trinh, 2017, p. 164).

The general exception clause is considered to be the clearest expression of States' intention to ensure that IIA obligations do not prevent the pursuit of public policies. However, it is still a matter of debate on whether the importation of general exceptions from trade agreements into investment treaties is an effective solution. First, some scholars argue that if the enforcement of an environmental measure meets the conditions of the exception clause (in a nondiscrimination, not arbitrarily, unreasonably manner, or without disguised barriers to investors), then such measure itself does not violate the principles of treatment in the IIA. Accordingly, it is unnecessary for the arbitral tribunal to apply to the exception clause (Newcombe and Paradell, 2009, p. 505). Secondly, this exception clause has not resolved the issue of compensation when the State enforces expropriation measures for public purposes—one of the most common claims in investor-State disputes in relation to the environment. In other words, the question remains as to whether this provision exempts the State from the obligation to compensate for the investor if the States to enforce an expropriation measure for public purposes under the scope of the exception clause. If not, such exemption might conflict with the fundamental purposes and principles of protection in international investment law, making investors bear all risks and thus threatening the development of international investment activities.

Unfortunately, these issues are left unanswered since the general exception clauses have only been included in some recent IIAs and not yet been interpreted by any arbitral tribunal in practice. In the WTO dispute settlement system, the general exception clause has been referred to in numerous disputes, which can be a useful source for investment arbitral tribunals. However, it should be noted that, in comparison with the WTO dispute settlement mechanism, the lack of an appellate body in the investor-State dispute settlement mechanism would result in the application and interpretation of general exception clauses being heavily dependent on each arbitral tribunal's perspectives.

## 6 Reserving Policy Space for Environmental Regulation for More Specific, Limited Subject Matters

A small group of IIAs concluded by Vietnam mentions the State's right to regulate with a limitation to more specific matters. The Morocco–Vietnam BIT (2012), for instance, at Article 2, mentions the State's right to take environmental measures as an exception to the principles of fair and equitable treatment (FET) and full protection and security (FPS). Paragraph 4 of this Article states that:

> This article [on FET and FPS] shall not apply to measures that have to be taken by either Contracting Party for reasons of public security, order or public health or protection of environment provided that such measures are not applied in a manner which would constitute a means of arbitrary or unjustified discrimination.

The specific areas where space for environment regulation is reserved also vary from treaties to treaties. It seems that Vietnam has not developed a consistent policy purpose to include such kind of language in IIAs, but more depends on other State partners. The CPTPP, focus on the provision of performance requirements, stating that:

> (d) Provided that such measures are not applied in an arbitrary or unjustifiable manner, or do not constitute a disguised restriction on international trade or investment", the provisions on Performance requirements shall not be construed to prevent a Party from "adopting or maintaining measures, including environmental measures:

(i) necessary to secure compliance with laws and regulations that are not inconsistent with this Agreement;
(ii) necessary to protect human, animal or plant life or health; or
(iii) related to the conservation of living or non-living exhaustible natural resources.

(Article 9.10. Performance requirements, Paragraph 3 (d))

Meanwhile the US–Vietnam Trade Relation Agreement 2000 even narrows the scope of policy space to environment regulations to the specific matters of technology transfer. Under Article 7 of this Treaty, the contracting Parties agree not to impose on covered investment any requirement to transfer technology except "when applying generally applicable environmental laws that are consistent with the provisions of this Agreement."

## 7 Indirect Expropriation

Despite the appearance of provisions to preserve the State's right to regulate environment issues, such measures are still exposed to claims of indirect expropriation, therefore resulting in the State's obligation to compensate affected investors. That is the reason why some countries negotiate to include environmental measures as an exception to the claims of indirect expropriation. Such kind of clause can be found in none of any BITs concluded by Vietnam, but in some TIPs. Analysis of these clauses also indicates significant similarity in the language. The ASEAN Comprehensive Investment Treaty (Annex 2 on Expropriation and Compensation, Paragraph 4.) and the AANZFTA (Annex on Expropriation and Compensation, Paragraph 4) preclude nondiscriminatory environment measures from the scope of indirect expropriation, providing that "Non-discriminatory measures of a Member State that are designed and applied to protect legitimate public welfare objectives, such as public health, safety and the environment, do not constitute an expropriation of the type referred to in sub-paragraph 2(b)."

In comparison, in the Korea–Vietnam FTA (2015), Annex 9-B. Expropriation, adds a proportional test to examine whether a nondiscriminatory regulatory action taken by the host State can be exempted from indirect expropriation:

> (ii) except in rare circumstances, such as, for example, when an action or a series of actions is extremely severe or disproportionate in light of its purpose or effect, non-discriminatory regulatory actions by a Party that are designed and applied to protect legitimate public welfare objectives, such as public health, safety, the environment, do not constitute indirect expropriations.

The proportionality test has been increasingly applied by Tribunals in a number of investor-State disputes when examining the provision of indirect expropriation, especially if the doctrine of police power is applied.[8] From a public policy perspective, the application of proportionality test can balance the environmental purpose of the measure with the degree of the measure's effect on the investor's right.

## 8 "Not Lowering Standards" Clause

"Not lowering standards" clause has been included in some new generation IIAs in order to restrict the government deregulation of environmental standards to attract or foreign investments (Camille, 2017, p. 542). This type of clause however rarely appears in the BITs of Vietnam. In fact, the Japan-Vietnam BIT (2003) is the only BIT with such language, stipulating that:

> The Contracting Parties recognize that it is inappropriate to encourage investment by investors of the other Contracting Party by relaxing environmental measures. To this effect each Contracting Party should not waive or otherwise derogate from such environmental measures as an encouragement for the establishment, acquisition or expansion in its Area of investments by investors of the other Contracting Party.

"Not lowering standards" clause was introduced due to the concern that some countries, because of pursuing investment policies, might lower environmental standards and create unfair competition among countries (Camille, 2017, p. 542). The problem, however, is that this type of provision does not directly resolve the conflict that exists in international investment law between lifting environmental standards and ensuring the principles of investor protection. In other words, this provision prevents countries from lowering environmental standards to attract investment, but fails to address the opposite question of whether a country can tighten its environmental measures even if that could cause difficulties to investment activities. These problems may explain why countries including Vietnam are not eager to use such provisions in their investment treaties.

---

[8] See Técnicas Medioambientales Tecmed, S.A. v. The United Mexican States, 2003, p. 46.

## 9 Environmental Matters and Investor-State Dispute Settlement

Given the fact that the existence of investors' right to bring a dispute against the host State poses major risks to the State's decision to take environmental measures, a possible resolution is to regulate environmental matters in the provision of investor-State dispute settlement. While investment arbitrators are often not experts in environment law, some IIAs include provision on the consultation of experts during the arbitral process. Vietnam has adopted this approach in several TIPs. A typical example of such clause can be found in Article 3.52 of the EVIPA which states that:

> The Tribunal, at the request of a disputing party or, after consulting the disputing parties, on its own initiative, may appoint one or more experts to report to it in writing on any factual issue concerning environmental, health, safety, or other matters raised by a disputing party in the proceedings.

A similar provision is also used in the ASEAN Comprehensive investment treaty. This provision is expected to bring other perspectives into the area of investment law, assisting the Tribunal to consider environmental issues in a more objective way. It, however, only deals with procedural issues rather than subjective matters. Further, it is at the Tribunal's discretion to assess the credibility of the expert reports and to decide how such reports may affect its decision.

In short, environmental issues have recently received increasing attention during the negotiation of IIAs. Although the approach might vary among IIAs, the general trend aims at ensuring a balance between commercial and noncommercial values, between State's right to regulate and commitments to protect investments. In other words, the development of international trade has shifted from a system that solely aims at protecting and encouraging investment to a system that pursues and promotes sustainable investment policies. Catching up with this trend, Vietnam has included a variety of provisions concerning the environment values in the recently-concluded IIAs. It is also noted that some IIAs of Vietnam such as the Japan–Vietnam BIT (2003) or EVIPA have used a combination of different types of environment language to deal with environment concerns. However, the survey also shows that the proportion of IIAs with the environment language of Vietnam is still very low as the majority of IIAs of Vietnam belong to the old-generation groups where the issue of environment was almost neglected. This poses a great risk to the country in the process of reforming the environment law system toward higher standards and stricter enforcement. Another problem is the lack of consistency in the way environment concerns are dealt with under different IIAs of Vietnam. Although it might be far from easy for Vietnam to adopt the same language when negotiating investment treaties with different nations, such considerable divergence may cause some challenges to the country in ensuring the compliance of environment measures with the obligations under different IIAs.

## 10 Conclusion

After more than 30 years of opening up to attract foreign investment, it is undeniable that foreign investment has contributed significantly to the development of Vietnam, stimulating investment resources, promoting economic growth, creating jobs, improving the quality of human resources and contributing to the country's greater integration into the global economy. However, besides the foregoing benefits, numerous FDI projects have caused serious environmental incidents that threaten the environment. Cases such as the untreated sewage of Vedan Company being released into Thi Vai River, that of Miwon Company being released to the Red River, or environmental disaster on the coast of the North Central region due to the discharge system of Ha Tinh Formosa Company—from Formosa Group (Taiwan) are not separate incidents but rather an alarming bell for the shortcomings of FDI in Vietnam (Hoang and Nguyen 2016) These shortcomings also set out challenges for the State in pursuing a more sustainable investment attraction policy, in harmony with the Government of Vietnam's view of "not trading environment for economic benefits" (Tuan 2019).

Facing the foregoing challenges, the promulgation and adjustment of laws as well as strengthening the management of FDI projects are of necessity to Vietnam. However, in this process, the State should pay attention to the commitments in the signed IIAs. As analyzed, environmental and public health measures if not being properly enforced, may trigger the violation of IIAs, exposing the State to the risk of being sued by investors. Thus, before enforcing measures for the purpose of protecting the environment and public health, it is recommended for Vietnam to cautiously assess the impact and ensure harmony between these measures and the State's obligations under international treaties.

The research on 67 BITs and 6 selected TIPs to which Vietnam is a contracting party shows that only few of them have addressed environmental issues. It should be noted that environmental concerns might appear in different parts of IIAs, for example, in the preamble, or briefly mentioned in the body text such as the general exception clause, the clause on "non-derogation of environmental standards," etc. Based on the review, it is also recommended that Vietnam should develop a negotiating plan to adjust the old-generation IIAs. For example, Vietnam and Finland once negotiated the signing of the 2008 Vietnam–Finland BIT to replace the old 1993 BIT, in which the role of environmental matters has been recognized at the preamble.

In addition, Vietnam is recommended to include environmental and public health provisions in the upcoming IIAs. Previously, the inclusion of environmental issues in investment treaties was often offered by developed countries; however, with regard to investment, it is developing countries like Vietnam to be more vulnerable. Therefore, it is reasonable for Vietnam to actively initiate the inclusion of such provision in IIAs, which is deemed a common trend in the drafting of IIAs.

In conclusion, environmental protection poses a great challenge to the principles of international investment law—a system designed to protect and promote economic objectives. Practice shows that the conflict between public interests and investors' interests exists in the majority of countries, and when dealing with such conflict, States often encounter the investor's opposition, or be brought before an international arbitral tribunal, or even incur compensation liabilities due to the breach of the IIA. Therefore, environmental concerns are now being expressly included in numerous IIAs, which once again emphasize the important role of these values in the investment policies of states. It is important for Vietnam, a country that is highly vulnerable to FDI's negative impacts, to seize this opportunity to make it possible to pursue a sustainable investment policy that remains consistent with its international commitments under investment treaties.

## References

### *Article/Books/Reports*


Benedetto SD (2013) International investment law and the environment. Edward Elgar Publishing

Bernasconi-Osterwalder N, Cosbey A, Johnson L, Vis-Dunbar D (2012) Investment treaties and why they matter to sustainable development: questions and answers. International Institute for Sustainable Development

Camille M (2017) Balancing investors' rights with environment protection in international investment arbitration: an assessment of recent trends in investment treaty drafting. International Lawyer 50(3):529–583

Dordi C, Nguyen TT (2017) Textbook on international investment law. Youth Publishing House, Hanoi

Gordon K, Pohl J (2011) Environmental concerns in international investment agreements: a survey. OECD Publishing

Gray KR (2002) Foreign direct investment and environmental impacts - is the debate over? Rev Eur Comp Int Environ Law 11(3):306–313. https://doi.org/10.1111/1467-9388.00329

Newcombe A, Paradell L (2009) Law and practice of investment treaty standards of treatment. Kluwer Law International

Nguyen TL (2018a) Một số tranh chấp liên quan đến biện pháp truất hữu trong quản lý môi trường—Kinh nghiệm phòng ngừa tranh chấp cho Việt Nam (Several disputes concerning expropriation in the environmental regulations—Experience to prevent disputes for Vietnam). In: Tran VD, Nguyen TL (eds) Giải quyết tranh chấp đầu tư quốc tế (International Investment Dispute Settlement). Ho Chi Minh National University Publisher, Ho Chi Minh, pp 121–144

Nguyen VT (2018b) Environment disputes in Vietnam: case studies from distributive and procedural justice perspectives. Vietnamese Law and Legal Forum Magazine

Slater TL (2015) Investor-state arbitration and domestic environmental protection. Wash Univ Global Study Law Rev 14(1):131–154

Tienhaara K (2006) What you don't know can hurt you: investor-state disputes and the protection of the environment in developing countries. Glob Environ Polit 6(4):73–100. https://doi.org/10.1162/glep.2006.6.4.73

Trinh HY (2017) Giáo trình luật đầu tư quốc tế (International Investment Law Textbook). National Political Publishing House, Hanoi

## Cases

Compañia del Desarrollo de Santa Elena S.A. v. Republic of Costa Rica, ARB/96/1 (ICSID February 17, 2000)
S.D Myers, Inc. v. Government of Canada (UNCITRAL Arbitration December 30, 2002)
Técnicas Medioambientales Tecmed, S.A. v. The United Mexican States, ARB (AF)/00/2 (ICSID May 29, 2003)

## Others

Hoang TH, Nguyen TD (2016, August 5) Bàn về vấn đề bảo vệ môi trường ở khu vực FDI (Discussion on environment protection in FDI areas). Retrieved February 15, 2020 from Môi trường và Cuộc Sống (Environment and Life): http://moitruong.net.vn/ban-ve-van-de-bao-ve-moi-truong-o-khu-vuc-fdi/
Tuan D (2019, December 31) Thủ tướng: "Không đánh đổi môi trường, văn hóa, văn minh xã hội để lấy kinh tế" (Prime Minister: "Not trading environmental, cultural and social interests for economic benefits"). Retrieved April 25, 2020 from Online newspaper of the Government: 2020: http://baochinhphu.vn/Hoat-dong-cua-lanh-dao-Dang-Nha-nuoc/Thu-tuong-Khong-danh-doi-moi-truong-van-hoa-van-minh-xa-hoi-de-lay-kinh-te/383913.vgp
UNCTAD (n.d.-a) Country navigator - Vietnam. Retrieved February 15, 2020 from Investment Policy Hub: https://investmentpolicy.unctad.org/country-navigator/237/viet-nam
UNCTAD (n.d.-b) Mapping of IIA content. Retrieved March 15, 2020 from Investment Policy Hub: https://investmentpolicy.unctad.org/international-investment-agreements/iia-mapping

**Kim Anh-Dao** is a law lecturer at Foreign Trade University. She teaches investment law, corporate law and international business law. Previously, she had worked as an in-house legal counsel in several leading multisector corporations in Vietnam such as Hoa Phat Group, SSI Securities Corporation, and VinGroup. She has legal and contracting experience through joining in various projects related to franchising contracts, e-commerce, real estates, and merger and acquisition transactions. She earned her bachelor degree in International Trade Law from Foreign Trade University, her master degree in Business Law from University of Adelaide (Australia) and her doctoral degree in International Law from Korea University (Korea). Her areas of interests include international investment law, commercial law, and dispute resolution.

# Barriers, Self, Society, and Sustainable Consumer Behavior in Vietnam

**Cam Anh-Thi Pham, Nhu-Quynh Dang, and Mai-Phuong Nguyen**

**Abstract** Understanding sustainable consumer behavior is central to a sustainable future and behavior change. Consumer behavior issues have been increasingly recognized by scientists and policymakers in terms of what drive consumers toward and what drive them away from sustainable consumption behavior, and how and how much people consume directly impacts the environment.

Vietnam is an emblematic new center of consumption in Asia and one of the centers of new middle-class consumers, giving us a lens through to view changing consumption patterns. In order to target the drive for sustainable development in Asia, Vietnam middle-class consumers have proven to be a critical group.

This chapter discusses drivers of sustainable consumer behavior, thus providing guidance for firms, policymakers, and consumers themselves to empower sustainability. The results of these studies were frequently inconclusive and sometimes contradictory. The second part is to focus more narrowly on consumers' motivations and psychological factors driving environmentally consequential behaviors. Areas of scientific inquiry that have dominated research agendas will be identified including cognitive barriers, consumers themselves, social influence, and product characteristics. Throughout the chapter, sustainable consumer behavior in Vietnam's context will be focused.

C. A.-T. Pham (✉)
Foreign Trade University, Ha Noi, Vietnam
e-mail: phamthicamanh@ftu.edu.vn

N.-Q. Dang
Academy of Policy and Development, Ha Noi, Vietnam
e-mail: nhudq@apd.edu.vn

M.-P. Nguyen
MobiFone Corporation, Ha Noi, Vietnam
e-mail: phuong.nguyenmai@mobifone.vn


H.-T. Vu, L.-P. Dana (eds.), *Vietnam on the Road to Sustainable Economic, Social and Business Development*, Frontiers in South and Southeast Asian Development Research, https://doi.org/10.1007/978-981-96-4371-4_8

# 1 Sustainable Consumer Behavior

## 1.1 *Roles of Consumers*

### 1.1.1 Why Do We Study the Roles of this Subject?

This section considers the roles which consumers play in promoting sustainable living. This chapter follows a marketing approach, investigating the roles of consumers in relation to social responsibility. Regarding this topic, Gupta and Singh (2017) proposed a twofold view: consumers as destroyers and developers. The consumers have to face the predicament when trying to maintain an easy, comforting means of living when it expresses chances to do harm to the environment.

### 1.1.2 Consumers Are Seen as Destroyers

There are statements that humans create impacts on the environment as a side-product, while pursuing comfort, labor reliefs, and many more. This motivation is not only applied to the consumption made by humans but also the acts done by intangible entities such as companies and technologies. Recently, there is an increasing trend in choosing sustainable consuming options over less-sustainable ones, both in day-to-day purchases and more high-value products purchasing. However, the emergence of the sustainable lifestyle trend is proposed to likely root from the concern to their own well-being rather than the environment. A key motive for behavior change is the cost factor, which could be clearly seen in the case of water, electricity, and fuel consumption. Stern (2000), by observing human behaviors in the marketplace, sums up that the intent behind the conversion to more efficient and environmentally friendly power source is the consumers' cost-saving benefits, not about their positive environmental contribution. Many behaviors and acts of consumption lead to them being seen as the destroyers of the environment.

### 1.1.3 Consumers Are Seen as Expected Developer

It is clear that humans have a great ability to make changes to the environment. It is also undeniable that one of the irreplaceable conditions for human survival on Earth is the establishment and maintenance of a clean and healthy environment. Due to human's capability of impacting the living environment, the solutions to environmental problems are within their power. The consumers, while being considered destroyers, are also expected to be developers. They are expected to balance between taking care of their personal welfare and making choices that are the best for the society they engage in, sometimes there are trade-offs to make. By discussing the essences of what is good for the public and the means to reach it, the consumers could sustainably develop their living world.

Consumers who are aware of the consequences of their consuming choices on the environment fall into a dilemma when their own well-being contradicts with the welfare of the environment. While the concerns for the environment often require consumers to put in efforts to reduce, reuse, and recycle, environmentally irresponsible use of environmental resources asks for no complication. An obvious example is in the case of transporting: it is a fact that the environment benefits from consumers sharing vehicles, leading to optimizing the routes, resulting in decreasing emission. However, to an individual, commuting on an owned car is often more comfortable than on a public-shared vehicle such as trains, buses, or trams. By making a more socially responsible choice of transporting via public means, consumers sacrifice the easiness and comfort that driving their personal cars offers. Such dilemma situations occur daily, and each of them confront socially concerned consumers with self-conflicts—when considering an easy solution that may harm the environment in short or long term, and a sustainable alternative. This divides consumers into two mainstreams. A group of people are willing to sacrifice the simplicity and comfort unsustainable choices come with in exchange of benefiting the environment or not harming it. Besides, some might totally ignore the effects of irresponsible consumption on sustaining the living environment.

### 1.1.4 Consumers Can Tremendously Impact the Environment

There exists a direct connection between consumption and sustainability. This connection is signaled by the fact that the world is experiencing many problems and human are to blame for most of them. A number of researches about weather conditions shows a close link of the increase of greenhouse gases and drastic changes in weather patterns. In fact, the upraising concentration of greenhouse gases is a significant sign which was recorded in recent years. Greenhouse gases consist of carbon dioxide, nitrous oxide, and methane. Human industrialization, urbanization, fossil fuels abusing, and automobile emissions have increased carbon dioxide concentration in the air by more than 30%. Nitrous oxide is largely released to the environment through the use of commercially produced fertilizers in the manufacturing chain of cotton and other activities in agriculture. Various aspects of human consumption that runs from the items being bought, the materials products are made, the degree of consumption to the disposal practices, all affect the probability of sustainability one product could perform. Humans are depicted as destroyers as well as expected developers of the environment because of the impactful human activities could put on the living environment. Thus, human consumption is strictly linked to sustainability.

# 2 Sustainable Consumer Behavior

## *2.1 Consumer Behaviors*

There are a variety of approaches to the definition of consumer behavior and one category: sustainable consumer behavior. There are some significant perspectives about sustainable consumer behavior, including the policy maker's view, the marketing view, the consumer interest focus, and the ethical focus.

In 1972, Bennet mentioned consumer behaviors as behaviors that consumers demonstrate in finding, buying, using, and evaluating products and services which were expected to satisfy their personal demands.

In 2008, Lamb, Hair, and McDaniel used the term "consumer behaviors" to suggest the decision-making process about choosing or eliminating certain types of products and services.

## *2.2 Sustainable Consumer Behaviors*

OECD defined sustainable development as the use of goods and services that satisfies, fulfills essential needs, and brings a higher quality of life, as long as decreasing the reliance and application of naturally sourced resources, lethal materials, lessening the releasing of harmful emissions through its life-circle while not negatively effect on the needs of future generations.

Sustainable consumer behavior could be seen as a subcategory of consumer behavior study that answers questions as why and how consumers decide to or not to consider sustainability when making, purchasing, and using goods and services. In detail, the study of sustainable consumer behaviors investigates the way consumers select, use, and dispose of products in regards to their personal sustainability goals. Researches in the field of consumer behaviors imply that the decision to choose, buy, consume, or deny a consumer product and service is largely influenced by the combination of external and internal forces. This knowledge covers the motivations which encourage people to behave sustainably.

Stating that the desire of most people is to get their present needs fulfilled without harming the environment, Stern (2000) characterized environmental and sustainable behavior by the degree the intention of benefiting or, at least, limiting negative impacts on the environment could influence the consumers' decisions. This approach considers consumer behavior study to fall under the umbrella of consumer psychology. Following this perspective, researchers have access to the reason behind environmentally concerned people's sustainable and unsustainable practices. The understanding of human behavior in consumption sheds light on how to cultivate a sustainable future and suggest practical ideas leading to phenomenally behavioral changes.

## 3 History of Research on Sustainable Consumer Behavior

At its inception, research on sustainable behavior focused on a concept termed "green consumers." This idea was identified and characterized as an attempt to evaluate its viability as a unique market segment, which is a defining activity of marketing. As a result, studies related to SB were directed at measuring sociopsychological and situational determinants of SCD such as demographic information, economic and cultural background, attitudes, and motivation, in correlation with measures of social consciousness and concerns with the environment. The work at this stage pointed out some key environment related behaviors such as recycling, environmental attitudes, and the willingness to adopt more environmentally friendly products. With the marketing foundation, early sustainability studies were often inconclusive and sometimes contradictory.

The ultimate finding was that personality variables were better predictors of ecological concern than socioeconomic demographic variables.

Conducting studies in this domain, researchers often struggle with the demand effects when consumers are motivated to give positive answers when asked while not maintaining or performing acts that concur with the responses expressed.

For example, in 1975, Webster stated that there were weak correlations of personality, attitudes, and other socioeconomic variables and disagreed with the conclusion of the relationship between previously used measures of social responsibility to socially conscious consumption.

In the 1980s, research in the domain of sustainability largely revolved around the key issue of energy conservation with some exceptions investigating the reactions of consumers to environmental legislation, specifically, recycling. Following the same viewpoint, the majority of studies in this period focused on individual attitudes and environmental beliefs in relation to behavioral intentions. Overall, these studies led to a similar conclusion of inconclusive and fragmented knowledge related to sustainable consumer behaviors. Some researches could be named:

In 1981, Leonard-Barton proved that there exists a link between voluntary simplifying lifestyle, which is indicated by behaviors of including environmental items, practicing recycling, composting, and travelling by bike to work, which implies energy-efficient behavior and high likelihood to make solar equipment purchases. Indicators as voluntarily simplifying style of living were considered attitudinal. Such indicators, however, were found to be poor and inconsistent predictors of sustainable energy conversion behaviors by the vast majority of the researches at the same time frame. For instance, Heslop et al., (1981) showed that price sensibility and energy use have a relationship; however, there was no connection between social responsibility or environmental awareness and the practice of saving energy. The minimal effects of attitudinal factors on behavioral changes were confirmed in the case of the study conducted by McNeill and Wilkie (1979). The research was about the value consumers perceived of the federal labeling program for home appliances. The study showed that consumers saw the information as useful and valuable, though not making influential impacts on their consuming choices.

In the 1990s, energy-saving and recycling kept on being the repeatedly mentioned topics of research, along with an uprising trend of "green" advertising. Continuing with a marketing approach, many researchers in this time period took a more macro-level look at sustainability and environmental consciousness, using them as the hearts of marketing strategies. In 1995, Shrum et al. created a survey at a large scale, attempting to draw a psychographic profile of green consumers to gain better understanding and produce recommendations for green marketing and advertising. In 1997, Menon and Menon first developed a framework that clarified a strategic management flow, integrating the existing concerns for the environment of the consumers with corporate social activities goals and corporate entrepreneurial strategy. The study conducted two important groups of factors that impact the adoption of green environmental entrepreneurial marketing strategy: the antecedents and the consequences. To be in detail, the antecedents included external policies, external economy, internal policies, and internal economy and the consequences consist of company performances and reputations. This new bright view on the appliance of sustainability, environmentalism in marketing suggested a way for much further development relating to corporate social responsibility and systems sustainability—a large field of investigation in later years.

In the 2000s, there emerged a drastic change in scholars' approaches to the topic. From 1970s to 1990s, the focal perspectives taken were clearly for the use of firms' marketing activities, or in general, to support a managerial purpose. Starting in the early 2000s, the perspectives taken on researching sustainable consumer behaviors shifted to decision-making. This approach focused more narrowly on the motivations and other factors that promoted consumers to perform environmentally consequential behaviors. Studies of this decade gave people insights as to what the drivers of consumers' environmentally minded behaviors are, both in the forms of descriptive understanding and formulated frameworks. The establishment of the four areas of research in consumer behavior studies showed a significant development of knowledge of this field through decades.

# 4 Drivers of Sustainable Consumer Behavior

## *4.1 Cognitive Barriers to Sustainability*

The cognitive processes underlying human decision-making present significant challenges to fostering sustainable behavior. Psychological research identifies two distinct mental systems that guide our choices. The first system, characterized by speed and intuition, operates through automatic, emotion-driven responses shaped by familiarity and instinct. In contrast, the second system is deliberate, reflective, and analytical, demanding conscious effort to process information and contemplate long-term outcomes (Sloman, 1996). These systems do not function in isolation; instead, they interact dynamically, with the rapid judgments of System 1 either endorsed or overridden by the deliberate reasoning of System 2 (Morewedge and Kahneman, 2010).

The complexity of sustainability decisions lies in their intrinsic demand for a forward-looking perspective, often requiring individuals to prioritize distant benefits over immediate convenience. For instance, the decision to commute via public transport rather than driving may initially evoke resistance—System 1 instinctively opposes the discomfort of waiting in the cold or the inconvenience of walking additional blocks. However, System 2 engages in a more comprehensive evaluation, recognizing the broader advantages, such as reduced traffic stress, improved physical health, and a significant reduction in carbon emissions. This process highlights the tension between the instant gratification favored by System 1 and the more effortful, environmentally conscious choices endorsed by System 2. Everyday scenarios, such as holding onto a recyclable item until finding an appropriate bin rather than discarding it in the nearest trash, exemplify this cognitive conflict (Brothers et al., 1994).

This interplay between short-term ease and long-term gain underscores one of the most profound barriers to sustainable behavior: the perception of effort and sacrifice. As Weber (2017) astutely observes, human decision-making is inherently biased toward preserving the status quo, often at the expense of future benefits. This present-focused cognitive bias hinders investments in sustainable products and practices, such as energy-efficient technologies, even when their long-term economic and environmental advantages far outweigh their initial costs (Gillingham et al., 2009). Addressing this challenge requires reimagining sustainability initiatives to make eco-friendly choices not only accessible but also seamlessly integrated into daily life. By reducing perceived effort and emphasizing immediate and tangible benefits, such strategies can bridge the divide between short-term convenience and long-term rewards, fostering a more sustainable future.

## 4.2 *The Self*

Understanding sustainable behavior requires examining how individuals perceive and express their self-identity. People often engage in environmentally friendly actions not just to align with their values but also to signal these values to themselves and others. This self-signaling can fulfill multiple motivations: reinforcing personal identity (Belk, 1988), gaining social recognition or status (Griskevicius et al., 2010), and projecting a positive image to others for reputational benefits (Semmann et al., 2005). These motivations frequently overlap, creating a complex foundation for sustainable behavior, where the act of choosing eco-friendly options becomes as much about self-validation as environmental impact.

Sustainability choices often involve a subtle psychological negotiation. When individuals select actions like turning off lights or reducing driving, they are not merely aiming for practical conservation but are also reaffirming their moral adequacy, a concept Steele (1988) describes as feeling "good and virtuous." Interestingly, research shows that people tend to prioritize personal behaviors over product-driven solutions, such as using energy-efficient appliances, when identifying effective environmental strategies (Attari et al., 2010). This preference highlights how deeply rooted moral evaluation is in sustainability, with emotions like pride and guilt playing significant roles in guiding behavior (Aquino and Reed, 2002).

Commitment and accountability can amplify this dynamic. For example, a large-scale experiment in a hotel setting found that guests who visibly committed to sustainable practices—such as wearing pins indicating their pledge—were more likely to conserve energy and reuse towels (Baca-Motes et al., 2013). These acts suggest that sustainability choices are not always driven by pure altruism but also serve as mechanisms to enhance self-concept. Moreover, when individuals perceive a threat to their self-identity, they often use eco-friendly actions to restore balance, such as choosing green products over superior alternatives, not just for their environmental benefits but to affirm personal values (Gao et al., 2009).

Paradoxically, the same self-signaling that motivates sustainable choices can sometimes lead to moral licensing. By reinforcing a virtuous self-image, initial eco-friendly actions may reduce the emotional barriers to unethical behavior. For instance, Mazar and Zhong (2010) found that consumers who purchased green products were more likely to engage in dishonest actions afterward, as their enhanced self-perception diminished guilt. Similarly, "vicarious moral cleansing" occurs when individuals associate themselves with socially responsible brands, allowing them to justify unethical behavior under the brand's moral umbrella (Newman and Brucks, 2018).

Sustainability decisions are also heavily influenced by emotional feedback. Recycling, for instance, not only alleviates guilt associated with waste but also generates pride from resourceful disposal (Sun and Trudel, 2016). However, the act of recycling can unintentionally encourage higher resource consumption, as individuals feel morally justified to use more when they know they can recycle (Catlin and Wang, 2013). These findings reveal that sustainable behavior is not merely a rational choice; it is a deeply emotional process intertwined with self-identity, social dynamics, and moral evaluation, challenging conventional approaches to promoting environmental responsibility.

## *4.3 Social Influence*

The adjustment of an individual's attitudes or behaviors as a result of the presence of others in a society is known as social influence. Social influence is the most effective way to influence sustainable customer behavior. Consumers are often affected by the appearance, behavior, and assumptions of others. One of the most important forces driving sustainable consumer behavior improvement is social factors (Abrahamse and Steg, 2013). In this section, we will look at how three distinct aspects of social power—social norms, social identities, and social desirability—can shift consumers to make more sustainable decisions.

## *4.4 Social Norms*

Social norms, which are unwritten guidelines that regulate social behavior and are formed and established by mutual interactions with a social community, are one of the most prevalent types of social influence. According to Cialdini (2004), Cialdini and Trost (1998), and Sunstein (1996), social norms reflect social approval and disapproval by deciding what should and should not be done; the consequences of breaking a social norm include peer disapproval and social prohibitions. Furthermore, social norms or beliefs on what is socially acceptable and acceptable in a certain situation can have a significant impact on sustainable customer behavior. As a result, there are social norms for almost any aspect of human activity (Cialdini, 1993).

Research across various domains consistently highlights the power of social norms in shaping environmentally responsible behavior. Social norms serve as a strong predictor for actions such as reducing litter (Cialdini et al., 1990), engaging in composting and recycling (Oskamp et al., 1991; White and Simpson 2013), and conserving energy (Goldstein et al., 2008; Schultz et al., 2007; Dwyer et al., 2015; Jachimowicz et al., 2018). They also influence consumer choices, such as selecting sustainably sourced food (Dowd and Burke 2013), adopting eco-friendly transportation options (Harland et al., 1999), and opting for green hotels (Teng et al., 2015).

Additionally, social norms impact decisions related to adopting renewable energy solutions like solar panels (Bollinger and Gillingham, 2012), adjusting energy consumption behaviors (Allcott and Mullainathan, 2010), and increasing participation in recycling programs (Meng and Trudel, 2017). Even seemingly small actions, such as reusing towels in hotels, are significantly influenced by the presence of social norms. These findings underscore the pivotal role that shared social expectations play in driving sustainable behaviors, suggesting that leveraging social influence can be a powerful strategy for promoting environmental responsibility.

The Theory of Planned Behavior has been applied to sustainable behaviors, and it assumes that, in addition to subjective norms, beliefs, and perceived behavioral influence shape intentions, which predict behavior (Han and Stoel, 2016; Heath and Gifford, 2002). This framework is used by Cialdini and his colleagues to define the expression "descriptive norm," which refers to knowledge on what other people are doing or commonly do (Cialdini et al., 1990; Reno et al., 1993). For example, normative data on the number of people taking action can be conveyed in marketing messages such as "90% of Boston residents recycle," "Two-thirds of your neighbors source their energy through renewable sources," or "Your energy consumption last month was above the average in your neighborhood." In the research of Goldstein et al., (2008), the utility of these types of informative norm messages is shown. Consumers who received descriptive norm messages in their hotel rooms ("Join your fellow guests in helping save the environment") were more likely to reuse their towels than those who received a regular environmental message ("Help save the environment"), according to two field experiments studying hotel towel reuse (energy conservation).

Descriptive norms, which reflect the common behaviors of others, often serve as stronger drivers of sustainable consumer actions than individual self-interest, yet their impact is frequently underestimated (Nolan et al., 2008). These norms become particularly compelling when they are tied to relatable contexts. For example, conveying that other hotel guests participated in an energy conservation program proved more effective than general environmental appeals. This effect was especially pronounced when the message referenced the specific hotel room the guest occupied, creating a personalized sense of relevance and social alignment (Goldstein et al., 2008). By demonstrating what others are doing in similar situations, descriptive norms provide a concrete and socially grounded framework for decision-making, enhancing their influence on sustainable behaviors (Fornara et al., 2011).

However, the power of descriptive norms can be a double-edged sword. When the majority of people are not engaging in the desired sustainable action, emphasizing such norms can unintentionally discourage participation, as it signals low prevalence of the behavior (Cialdini, 2003; Schultz et al., 2007). On the other hand, positive modeling can amplify their effectiveness. This finding highlights the critical role of strategic framing in leveraging descriptive norms. By carefully presenting these norms to emphasize positive participation, they can effectively mobilize communities and encourage widespread adoption of sustainable practices.

Regardless, "injunctive norms" express what behaviors other people approve of and disapprove of. Such norms can thus affect sustainable behavior, but they must be used with caution (Kronrod et al., 2012). According to Kronrod et al., (2012), assertive, injunctive comments that portrayed sustainable behavior as a must, such as "Reducing air pollution: everyone must use more public transportation!" resulted in customer reactance to the issue. Injunctive norms are most effective when they are combined with thoughts about the ingroup and do not threaten feelings of autonomy, which can lead to "reactance" reactions. Consumers reported that their intention to adhere to the assertive message is lower, which depends on how important they discerned the matter to be, but replied positively to assertive messages when they believed the importance of the environmental issue. However, when they believed the environmental issue to be less important, suggestive messages were more effective than assertive messages.

Overall, research suggests that social influence, especially social norms, has a strong, predictable, and widespread influence on sustainable behaviors. Thus, both descriptive and injunctive norms can affect sustainable customer behaviors, but they should be used with care to avoid causing negative social reaction.

## *4.5 Social Identities*

In addition to individual identity, sustainable customer behavior is affected by social or group identification. Social groups which people belong to have impacted on people's attitudes, beliefs, and behaviors, according to the research conducted by Tajfel (1982), at times even more so than their personal identities (Onorato and

Turner, 2004). People are more likely to imitate habits that reinforce group identity and relations. In one example, climate change is a hotly debated topic that is being increasingly politicized. As a result, many of the gaps in views and opinions on climate change are partisan; conservatives and Republicans are less inclined than liberals and Democrats to accept the scientific facts and to express alarm about climate change (Brulle et al., 2012; Hardisty et al., 2010; McCright and Dunlap, 2011).

The attitudes and behaviors of individuals are profoundly shaped by the opinions and actions of those they identify with as part of their social group. When people value their membership within an ingroup, such as a political party, they tend to align their beliefs and actions with the group to reinforce their sense of belonging. Cohen (2003) demonstrated that political affiliations can significantly influence behavior; individuals often adopt their party's stance, regardless of the accuracy of the information, as a means of strengthening their group identity. This phenomenon is especially evident in reactions to climate science, where people's perceptions are often guided by the expressed attitudes of their political leaders. Given the complexity of climate issues, many individuals lack the expertise to independently evaluate the information and instead rely on cues from their ingroup for guidance.

Shifts in public opinion on climate change further illustrate the influence of political group dynamics. A 1997 Gallup poll revealed relatively similar levels of belief in climate change among Democrats (47%) and Republicans (46%). However, by 2008, the gap had widened dramatically, with 76% of Democrats and only 41% of Republicans expressing confidence in climate change. This divergence reflects the growing influence of political elites in shaping the narratives and beliefs within their respective groups (Dunlap, 2008; Pearson et al., 2016). These findings underscore the powerful role of social identity and group allegiance in shaping opinions, highlighting the need to consider ingroup dynamics when addressing public attitudes toward complex scientific issues.

Understanding the normative frameworks of different groups is vital for influencing consumer behavior and decision-making in sustainability. Markowitz and Shariff (2012) attributed divergent views on climate change to the distinct moral priorities of liberals and conservatives. Conservatives emphasize group cohesion, respect for authority, and sanctity, whereas liberals prioritize fairness and reducing harm. As climate change is often framed as a matter of justice and the harm it imposes on current and future generations, it tends to resonate more strongly with liberal moral values, leaving conservatives less engaged with the issue. Building on this, Kidwell et al., (2013) demonstrated that tailored advertising strategies based on political ideologies could encourage sustainable behaviors. Appeals crafted for liberals focused on empathy and fairness, while messages for conservatives emphasized responsibility and respect for authority. When these campaigns aligned with the audience's ideological values, they successfully promoted sustainable actions. However, messages based on opposing values were ineffective, highlighting the importance of matching sustainability messages to consumers' political philosophies.

Social identity strongly influences behavior, with individuals often avoiding actions or products that could jeopardize their group affiliation. People may deliberately distance themselves from behaviors associated with dissociative reference groups, which are groups they wish to avoid. White et al., (2014) found that when individuals realized a dissociative group had outperformed them in sustainable practices, they actively improved their own group's performance and adopted more sustainable behaviors. This competitive motivation underscores how group dynamics can drive environmentally friendly actions.

The pursuit of status is a deep-rooted human motive, historically tied to lineage and, more recently, to wealth and material possessions (Veblen, 1899; Han et al., 2010). Today, status is inferred from people's ability to acquire and sacrifice, often reflected in the brands they own and the lifestyle choices they make (Belk, 1988). Costly signaling theory explains how individuals use expensive or resource-intensive behaviors to signal their status and values. These actions, while seemingly wasteful or arbitrary, convey valuable information to others, demonstrating the individual's capacity to bear costs or make sacrifices (Bénabou and Tirole, 2006; Gneezy et al., 2012; Zahavi, 1975). In the context of sustainability, this theory suggests that visible eco-friendly choices, particularly those involving significant effort or expense, can serve as status signals, enhancing the individual's social standing while promoting sustainable norms.

Purchasing environmentally friendly products often acts as a public display of one's commitment to sustainability and social values. Griskevicius et al., (2010) describe green goods as "costly signals" because they typically involve higher prices and, at times, lower functionality compared to nongreen alternatives. This choice reflects a willingness to take on potential drawbacks in the interest of benefiting society, serving as a marker of both altruism and social responsibility.

In another context, Milinski et al., (2006) investigated collective action in a climate change public goods experiment. Participants were asked to allocate group profits to a shared climate fund, which was used to disseminate scientific knowledge on climate issues for the benefit of all. The fund's effectiveness was directly tied to the contributions made by the group. The study revealed that individuals were willing to sacrifice personal financial benefits to support societal goals, driven partly by a sense of altruism. Interestingly, the prospect of gaining social recognition or prestige further motivated participation, suggesting that the interplay between selflessness and status-seeking can significantly enhance engagement in sustainability initiatives. This dynamic highlights how personal and collective interests can align, promoting behaviors that benefit both society and the individual.

## 4.6 Social Desirability

The concept of "social desirability" plays a significant role in shaping sustainable consumer behavior, as individuals often make choices to create a favorable impression on others. Green and Peloza (2014) found that many consumers opt for environmentally friendly products to demonstrate positive social values, with high-involvement items like hybrid cars being used as symbols of status and commitment to sustainability (Griskevicius et al., 2010). However, this dynamic can be complex. Observers may sometimes perceive sustainable behaviors negatively, discouraging some individuals from adopting eco-friendly actions. For example, men have avoided engaging in environmentally conscious behaviors when such actions were stereotypically associated with feminine traits (Brough et al., 2016). This underscores the importance of framing sustainability in ways that enhance its social appeal while minimizing any negative stereotypes or misconceptions (Sadalla and Krull, 1995; Minson and Monin, 2012; Shang and Peloza, 2016).

Public settings further amplify the impact of social influence, as individuals tend to align their behaviors with what is deemed socially appropriate when they know they are being observed (Green and Peloza, 2014; Grolleau et al., 2009). Public commitments can also strengthen sustainable behaviors. For instance, those who made explicit commitments to participate in hotel energy conservation programs and visibly displayed their involvement, such as wearing pins, were more likely to follow through with their actions (Baca-Motes et al., 2013). By encouraging public acknowledgment and commitment, organizations can effectively leverage social dynamics to promote eco-friendly choices, fostering greater participation in sustainability initiatives.

## 5 Product

The appeal of sustainable products lies in their potential to benefit both the environment and society. However, consumer preferences for these products are influenced by various factors. While many individuals value eco-friendly features (Trudel and Cotte, 2009), their decision-making is often shaped by how options are presented (Irwin and Naylor, 2009), their personal beliefs and values (White et al., 2012), and the benefits they associate with specific product categories (Luchs et al., 2010). For instance, consumers are more likely to consider sustainable goods when presented with a diverse array of options (Broniarczyk et al., 1998; Whitley et al., 2018). However, the process of narrowing down choices plays a critical role, when consumers exclude options rather than select favorites, they are more likely to emphasize ethical considerations (Irwin and Naylor, 2009). Moreover, consumer expectations differ depending on the product type. Organic items, for example, are often linked to qualities like gentleness, making them attractive for certain uses, such as baby care, while conventional products are associated with strength and durability, which may overshadow sustainability in categories like automotive supplies (Luchs et al., 2010).

Product and packaging design significantly influence whether items are recycled or discarded. Research by Trudel and Argo (2013) and Trudel et al., (2016) shows that products undergoing visible physical changes during use—such as being crumpled, torn, or resized—are more likely to be thrown away than recycled. For instance, intact paper is far more likely to end up in a recycling bin than shredded or damaged paper. These findings underscore the importance of creating designs that minimize distortion. Packaging that remains intact, such as easy-to-open designs, can boost recycling rates by avoiding damage during use (Baxter and Childs, 2017). Similarly, durable materials that maintain their form and usability over time increase the likelihood of proper disposal. By aligning design with recycling behavior, manufacturers can encourage environmentally responsible choices, reducing waste and supporting sustainable practices.

## 6 Vietnam Context and Implications

Vietnam's rapid economic growth has led to a burgeoning middle class, significantly influencing consumption patterns. However, the shift toward sustainable consumer behavior faces several barriers. A study on green consumption behaviors in Vietnam highlights that, despite increased environmental awareness, significant obstacles hinder consumer engagement, including limited availability of green products and lack of trust in product claims (Trong Nguyen et al., 2023). Additionally, the higher cost of sustainable products and limited access to information further impede the adoption of eco-friendly consumption habits.

Social influences play a crucial role in shaping consumer behavior in Vietnam. Research focusing on Generation Z in Ho Chi Minh City indicates that social influence, along with environmental attitudes and pricing, significantly affects sustainable consumption intentions (Le et al., 2022). This underscores the importance of leveraging social networks and community norms to promote sustainable consumption. Policymakers and businesses can harness these insights by implementing targeted educational campaigns and creating supportive environments that encourage sustainable choices, thereby fostering a culture of sustainability among Vietnamese consumers.

## References

Abrahamse W, Steg L (2013) Social influence approaches to encourage resource conservation: a meta-analysis. Glob Environ Chang 23(6):1773–1785. https://doi.org/10.1016/j.gloenvcha.2013.07.029

Allcott H, Mullainathan S (2010) Behavior and energy policy. Science 327(5970):1204–1205. https://doi.org/10.1126/science.1180775

Aquino K, Reed A (2002) The self-importance of moral identity. J Pers Soc Psychol 83(6):1423–1440. https://doi.org/10.1037/0022-3514.83.6.1423

Attari SZ, DeKay ML, Davidson CI, Bruine de Bruin W (2010) Public perceptions of energy consumption and savings. Proc Natl Acad Sci 107(37):16054–16059

Baca-Motes K, Brown A, Gneezy A, Keenan EA, Nelson LD (2013) Commitment and behavior change: evidence from the field. J Consum Res 39(5):1070–1084. https://doi.org/10.1086/667226

Baxter W, Childs PR (2017) Designing circular possessions. Routledge, pp 391–404. https://doi.org/10.4324/9781315693309.ch26

Belk RW (1988) Possessions and the extended self. J Consum Res 15(2):139–168. https://www.jstor.org/stable/2489522

Bénabou R, Tirole J (2006) Incentives and prosocial behavior. Am Econ Rev 96(5):1652–1678. https://doi.org/10.1257/aer.96.5.1652

Bennett PD, Kassarjian HH (1972) Consumer behavior. Prentice-Hall, Englewood Cliffs

Bollinger B, Gillingham K (2012) Peer effects in the diffusion of solar photovoltaic panels. Mark Sci 31(6):900–912. https://doi.org/10.1287/mksc.1120.0727

Broniarczyk SM, Hoyer WD, McAlister L (1998) Consumers' perceptions of the assortment offered in a grocery category: the impact of item reduction. J Mark Res 35(2):166–176. https://doi.org/10.1177/002224379803500203

Brothers KJ, Krantz PJ, McClannahan LE (1994) Office paper recycling: a function of container proximity. J Appl Behav Anal 27(1):153–160. https://doi.org/10.1901/jaba.1994.27-153

Brough AR, Wilkie JEB, Ma J, Isaac MS, Gal D (2016) Is eco-friendly unmanly? The green-feminine stereotype and its effect on sustainable consumption. J Consum Res 43(4):567–582. https://doi.org/10.1093/jcr/ucw044

Brulle RJ, Carmichael J, Jenkins JC (2012) Shifting public opinion on climate change: an empirical assessment of factors influencing concern over climate change in the U.S., 2002–2010. Clim Chang 114(2):169–188. https://doi.org/10.1007/s10584-012-0403-y

Catlin JR, Wang Y (2013) Recycling gone bad: when the option to recycle increases resource consumption. J Consum Psychol 23(1):122–127. https://doi.org/10.1016/j.jcps.2012.04.001

Cialdini RB (1993) The psychology of influence. William Morrow & Co, New York

Cialdini RB (2003) Crafting normative messages to protect the environment. Curr Dir Psychol Sci 12(4):105–109. https://doi.org/10.1111/1467-8721.01242

Cialdini R (2004) Influence: science and practice. Allyn & Bacon, Boston

Cialdini RB, Reno RR, Kallgren CA (1990) A focus theory of normative conduct: recycling the concept of norms to reduce littering in public places. J Pers Soc Psychol 58(6):1015–1026. https://doi.org/10.1037/0022-3514.58.6.1015

Cialdini RB, Trost MR (1998) Social influence: Social norms, conformity and compliance. In Gilbert DT, Fiske ST, Lindzey G (Eds), The handbook of social psychology (4th ed., pp. 151–192). McGraw-Hill

Cohen GL (2003) Party over policy: the dominating impact of group influence on political beliefs. J Pers Soc Psychol 85(5):808–822. https://doi.org/10.1037/0022-3514.85.5.808

Dowd K, Burke KJ (2013) The influence of ethical values and food choice motivations on intentions to purchase sustainably sourced foods. Appetite 69:137–144. https://doi.org/10.1016/j.appet.2013.05.024

Dunlap RE (2008) Climate-change views: Republican-Democratic gaps expand. Retrieved from https://news.gallup.com/poll/107569/climatechange-views-republicandemocratic-gaps-expand.aspx

Dwyer PC, Maki A, Rothman AJ (2015) Promoting energy conservation behavior in public settings: the influence of social norms and personal responsibility. J Environ Psychol 41:30–34. https://doi.org/10.1016/j.jenvp.2014.11.002

Fornara F, Carrus G, Passafaro P, Bonnes M (2011) Distinguishing the sources of normative influence on proenvironmental behaviors. Group Process Intergroup Relat 14(5):623–635. https://doi.org/10.1177/1368430211408149

Gao L, Wheeler SC, Shiv B (2009) The "shaken self": product choices as a means of restoring self-view confidence. J Consum Res 36(1):29–38. https://doi.org/10.1086/596028

Gillingham K, Newell R G, Palmer K (2009) Energy efficiency economics and policy. Annu Rev Resour Econ 1(1):597–620

Gneezy A, Imas A, Brown A, Nelson LD, Norton MI (2012) Paying to be nice: consistency and costly prosocial behavior. Manag Sci 58(1):179–187. https://doi.org/10.1287/mnsc.1110.1437

Goldstein NJ, Cialdini RB, Griskevicius V (2008) A room with a viewpoint: using social norms to motivate environmental conservation in hotels. J Consum Res 35(3):472–482. https://doi.org/10.1086/586910

Green T, Peloza J (2014) Finding the right shade of green: the effect of advertising appeal type on environmentally friendly consumption. J Advert 43(2):128–141. https://doi.org/10.1080/00913367.2013.834805

Griskevicius V, Tybur JM, Van den Bergh B (2010) Going green to be seen: status, reputation, and conspicuous conservation. J Pers Soc Psychol 98(3):392–404. https://doi.org/10.1037/a0017346

Grolleau G, Ibanez L, Mzoughi N (2009) Too much of a good thing? Why altruism can harm the environment? Ecol Econ 68(7):2145–2149. https://doi.org/10.1016/j.ecolecon.2009.02.020

Gupta H, Singh S (2017) Sustainable practices through green guerrilla marketing—an innovative approach. J Innov Sustain RISUS 8(2):61. https://doi.org/10.24212/2179-3565.2017v8i2p61-78

Han TI, Stoel L (2016) Explaining socially responsible consumer behavior: a meta-analytic review of theory of planned behavior. J Int Consum Mark 29(2):91–103. https://doi.org/10.1080/08961530.2016.1251870

Han YJ, Nunes JC, Drèze X (2010) Signaling status with luxury goods: the role of brand prominence. J Mark 74(4):15–30

Hardisty DJ, Johnson EJ, Weber EU (2010) A dirty word or a dirty world? Attribute framing, political affiliation, and query theory. Psychol Sci 21(1):86–92

Harland P, Staats H, Wilke HA (1999) Explaining proenvironmental intention and behavior by personal norms and the Theory of Planned Behavior 1. J Appl Soc Psychol 29(12):2505–2528

Heath Y, Gifford R (2002) Extending the theory of planned behavior: predicting the use of public transportation 1. J Appl Soc Psychol 32(10):2154–2189

Heslop LA, Moran L, Cousineau A (1981) "Consciousness" in energy conservation behavior: an exploratory study. J Consum Res 8(3):299. https://doi.org/10.1086/208868

Irwin JR, Naylor RW (2009) Ethical decisions and response mode compatibility: weighting of ethical attributes in consideration sets formed by excluding versus including product alternatives. J Mark Res 46(2):234–246

Jachimowicz JM, Hauser OP, O'Brien JD, Sherman E, Galinsky AD (2018) The critical role of second-order normative beliefs in predicting energy conservation. Nat Hum Behav 2(10):757–764

Kidwell B, Farmer A, Hardesty DM (2013) Getting liberals and conservatives to go green: political ideology and congruent appeals. J Consum Res 40(2):350–367

Kronrod A, Grinstein A, Wathieu L (2012) Go green! Should environmental messages be so assertive? J Mark 76(1):95–102

Lamb CW, Hair JF, McDaniel C (2008) Marketing, 10th edn. Cengage Learning, Mason

Le TD, Duc Tran H, Hoang TQH (2022) Ethically minded consumer behavior of generation Z in Vietnam: the impact of socialization agents and environmental concern. Cogent Bus Manag 9(1):2102124

Leonard-Barton D (1981) Voluntary simplicity lifestyles and energy conservation. J Consum Res 8(3):243. https://doi.org/10.1086/208861

Luchs MG, Naylor RW, Irwin JR, Raghunathan R (2010) The sustainability liability: potential negative effects of ethicality on product preference. J Mark 74(5):18–31

Markowitz EM, Shariff AF (2012) Climate change and moral judgement. Nat Clim Chang 2(4):243–247

Mazar N, Zhong C-B (2010) Do green products make us better people? Psychol Sci 21(4):494–498. https://doi.org/10.1177/0956797610363538

McCright AM, Dunlap RE (2011) The politicization of climate change and polarization in the American public's views of global warming, 2001–2010. Sociol Q 52(2):155–194

McNeill DL, Wilkie WL (1979) Public policy and consumer information: impact of the new energy labels. J Consum Res 6(1):1–11

Meng MD, Trudel R (2017) Using emoticons to encourage students to recycle. J Environ Educ 48(3):196–204
Menon A, Menon A (1997) Enviropreneurial marketing strategy: the emergence of corporate environmentalism as market strategy. J Mark 61(1):51–67. https://doi.org/10.1177/002224299706100105
Milinski M, Semmann D, Krambeck HJ, Marotzke J (2006) Stabilizing the Earth's climate is not a losing game: supporting evidence from public goods experiments. Proc Natl Acad Sci 103(11):3994–3998
Minson JA, Monin B (2012) Do-gooder derogation: disparaging morally motivated minorities to defuse anticipated reproach. Soc Psychol Personal Sci 3(2):200–207
Morewedge CK, Kahneman D (2010) Associative processes in intuitive judgment. Trends Cogn Sci 14(10):435–440. https://doi.org/10.1016/j.tics.2010.07.004
Newman KP, Brucks M (2018) The influence of corporate social responsibility efforts on the moral behavior of high self-brand overlap consumers. J Consum Psychol 28(2):253–271. https://doi.org/10.1002/jcpy.1027
Nolan JM, Schultz PW, Cialdini RB, Goldstein NJ, Griskevicius V (2008) Normative social influence is underdetected. Personal Soc Psychol Bull 34(7):913–923
Onorato RS, Turner JC (2004) Fluidity in the self-concept: the shift from personal to social identity. Eur J Soc Psychol 34(3):257–278
Oskamp S, Harrington MJ, Edwards TC, Sherwood DL, Okuda SM, Swanson DC (1991) Factors influencing household recycling behavior. Environ Behav 23(4):494–519
Pearson AR, Schuldt JP, Romero-Canyas R (2016) Social climate science: a new vista for psychological science. Perspect Psychol Sci 11(5):632–650
Reno RR, Cialdini RB, Kallgren CA (1993) The transsituational influence of social norms. J Pers Soc Psychol 64(1):104
Sadalla EK, Krull JL (1995) Self-presentational barriers to resource conservation. Environ Behav 27(3):328–353
Schultz PW, Nolan JM, Cialdini RB, Goldstein NJ, Griskevicius V (2007) The constructive, destructive, and reconstructive power of social norms. Psychol Sci 18(5):429–434
Semmann D, Krambeck H, Milinski M (2005) Reputation is valuable within and outside ones? Own social group. Behav Ecol Sociobiol 57(6):611–616. https://doi.org/10.1007/s00265-004-0885-3
Shang J, Peloza J (2016) Can "real" men consume ethically? How ethical consumption leads to unintended observer inference. J Bus Ethics 139:129–145
Sloman. (1996) The empirical case for two systems of reasoning. Psychol Bull 119(22):3. https://doi.org/10.1037/0033-2909.119.1.3
Steele CM (1988) The psychology of self-affirmation: sustaining the integrity of the self. Adv Exp Soc Psychol 21(1):261–302. https://doi.org/10.1016/s0065-2601(08)60229-4
Stern PC (2000) New environmental theories: toward a coherent theory of environmentally significant behavior. J Soc Issues 56(3):407–424. https://doi.org/10.1111/0022-4537.00175
Sun M, Trudel R (2016) The effect of recycling versus trashing on consumption: theory and experimental evidence. SSRN Electron J. https://doi.org/10.2139/ssrn.3036490
Sunstein CR (1996) Social norms and social roles. Col Lit Rev 96:903
Tajfel H (1982) Social psychology of intergroup relations. Annual Review of Psychology, 33:1–39
Teng Y-M, Wu K-S, Liu H-H (2015) Integrating altruism and the theory of planned behavior to predict patronage intention of a green hotel. J Hosp Tourism Res 39(3):299–315
Trong Nguyen L, Nguyen TH, Ngoc Nguyen H, Dai Nguyen L, Thi Thu Nguyen D, Duy Le L (2023) Determinants of green consumer behavior: a case study from Vietnam. Cog Bus Manag 10(1):2197673
Trudel R, Argo JJ (2013) The effect of product size and form distortion on consumer recycling behavior. J Consum Res 40(4):632–643
Trudel R, Cotte J (2009) Does it pay to be good? MIT Sloan Manag Rev 50(2):61

Trudel R, Argo JJ, Meng MD (2016) Trash or recycle? How product distortion leads to categorization error during disposal. Environ Behav 48(7):966–985

Veblen T (1899) The theory of the leisure class—an economic study of institutions, reprint. The New American Library of World Literature, New York

Weber EU (2017) Breaking cognitive barriers to a sustainable future. Nat Hum Behav 1(1):13. https://doi.org/10.1038/s41562-016-0013

Webster FE Jr (1975) Determining the characteristics of the socially conscious consumer. J Consum Res 2(3):188. https://doi.org/10.1086/208631

White K, Simpson B (2013) When do (and don't) normative appeals influence sustainable consumer behaviors? J Mark 77(2):78–95

White K, MacDonnell R, Ellard JH (2012) Belief in a just world: consumer intentions and behaviors toward ethical products. J Mark 76(1):103–118

White K, Simpson B, Argo JJ (2014) The motivating role of dissociative out-groups in encouraging positive consumer behaviors. J Mark Res 51(4):433–447

Whitley SC, Trudel R, Kurt D (2018) The influence of purchase motivation on perceived preference uniqueness and assortment size choice. J Consum Res 45(4):710–724

Zahavi A (1975) Mate selection—a selection for a handicap. J Theor Biol 53(1):205–214

**Cam Anh-Thi Pham** (PhD.) got her doctoral degree in Public Policy at the University of Texas at Dallas, the United States in 2017. She is currently a lecturer of the School of Economics and International Business, Foreign Trade University. Her research interests include sustainable development, policy analysis, and international business.

**Nhu-Quynh Dang** (MSc.) for her master degree in International Management at the University of Sunderland, United Kingdom in 2018. She is currently lecturing at the International School of Economics and Finance, Academy of Policy and Development in Hanoi, Vietnam.

**Mai-Phuong Nguyen** (MSc.) graduated in International Economics from Foreign Trade University. She is currently working at MobiFone Digital Services, which is a branch of MobiFone Corporation, Vietnam. Her research interests span a wide range from customer behavior to corporate social responsibility and sustainable development.

# The Research Trends of Sustainability Finance in Vietnam

**Huong-Thi Thu Le and Ly-Thi Bui**

**Abstract** The emerging field of sustainable finance offers a transformative opportunity for Vietnam's financial system to contribute to a more sustainable future. Despite its importance, research on sustainable finance, particularly within the Vietnamese context, remains limited. This chapter examines research trends in sustainable finance, focusing on Vietnam while drawing comparisons to global developments, using text-mining methods to identify key themes and gaps in the literature. The chapter begins by reviewing widely used definitions and activities of sustainable finance, including sustainable funds, green bonds, and sustainability bonds, and proposes a precise and standardized definition tailored to Vietnam's financial landscape. It then explores critical areas requiring further research, such as the risks posed by climate change and biodiversity loss, the resilience of ESG frameworks, the impact of sustainable finance initiatives in Vietnam, and practical tools for building a more sustainable financial market. Finally, the chapter provides recommendations to guide future qualitative research on sustainable finance, emphasizing Vietnam's unique challenges and opportunities. By doing so, it aims to advance understanding and foster the development of sustainability-focused financial practices within Vietnam and beyond.

## 1 Introduction

Financial markets have become pivotal drivers of Vietnam's economic growth in the era of globalization and integration. Beyond their economic role, these markets are increasingly recognized for their potential to promote investments that benefit

H.-T. T. Le (✉)
University of Alberta, Alberta, Canada
e-mail: huong1@ualberta.ca

L.-T. Bui
Foreign Trade University, Dong Da, Ha Noi, Vietnam
e-mail: lybt@ftu.edu.vn

 
H.-T. Vu, L.-P. Dana (eds.), *Vietnam on the Road to Sustainable Economic, Social and Business Development*, Frontiers in South and Southeast Asian Development Research, https://doi.org/10.1007/978-981-96-4371-4_9

natural ecosystems and enhance social welfare. Vietnam, like the rest of the world, has witnessed significant changes in the financial landscape over the past decades. Since the global financial crisis in 2008, there has been a growing shift in focus from traditional financial theories to approaches that incorporate sustainability considerations. The coronavirus pandemic further underscored the vulnerabilities of financial systems and emphasized the urgent need to rebuild economies with resilience and sustainability at their core.

In Vietnam, sustainable finance has emerged as a critical pathway to achieving long-term economic and social objectives. This approach emphasizes the integration of environmental, social, and governance (ESG) factors into financial decision-making. Many researchers argue that sustainable finance plays a vital role in advancing financial theories by addressing the limitations of traditional models, which often prioritize short-term financial returns over broader societal and environmental outcomes. Instead, sustainable finance broadens the scope to include intergenerational equity, ecological resilience, and social well-being, making it highly relevant to Vietnam's development goals.

Vietnam's unique economic and environmental challenges demand innovative approaches to sustainable finance. With fewer restrictive assumptions and greater applicability, sustainable finance offers a robust framework for addressing issues such as climate change, urbanization, and the need for equitable development. However, transforming Vietnam's financial system into a more sustainable model requires further research and strategic guidance.

This chapter explores the research trends in sustainable finance specific to Vietnam, providing insights into emerging priorities and opportunities for scholars and policymakers. It leverages text-mining techniques to identify key areas of interest and gaps in the existing literature. Sustainable finance in Vietnam is defined here as any financial service that integrates ESG criteria to achieve long-term benefits for customers, businesses, and society. This includes activities such as green bonds, impact investing, microfinance, sustainable funds, and credits for environmental projects.

The chapter aims to refine the understanding of sustainable finance within Vietnam's context, outline its key tools, and highlight future research directions. It identifies Vietnam's potential to enhance economic efficiency, support ecological preservation, and foster cultural and social diversity through sustainable financial practices. By focusing on Vietnam, this chapter emphasizes the importance of aligning global sustainability principles with local development needs.

The chapter is organized into five sections. Section 2 provides a comprehensive review of the literature on sustainable finance in Vietnam. Section 3 compares recent studies on the subject, highlighting Vietnam's unique contributions and challenges. Section 5 examines emerging research trends in sustainable finance, with a focus on Vietnam's economic and social priorities. Finally, Sect. 5 concludes with recommendations for future research and practical strategies to strengthen Vietnam's sustainable financial systems.

## 2 Literature Review

### 2.1 *Definition of Sustainability Finance*

Sustainable finance is an emerging field that plays a critical role in building a sustainable future. However, the terminology surrounding sustainable finance can be perplexing, as few standard definitions exist, and different nations and organizations often use the same terms to describe varying concepts. Establishing a precise definition of sustainable finance is essential for scholars and financial institutions to advance research and practice in this field.

The European Commission defines sustainable finance as "any form of financial service integrating environmental, social, and governance (ESG) criteria into business or investment decisions for the long-lasting benefit of both customers and society at large" (EU, 2020). This encompasses environmental considerations such as climate change mitigation, biodiversity preservation, and pollution prevention; social considerations like addressing inequality, promoting inclusiveness, and ensuring human rights; and governance factors that ensure responsible decision-making within institutions.

In the European policy context, sustainable finance is seen as a means to foster economic growth while addressing environmental and social challenges. It emphasizes transparency regarding ESG-related risks and aims to mitigate these risks through robust governance. The European Commission positions sustainable finance as a tool to mobilize private investments toward achieving the European Green Deal's objectives, transitioning to a climate-neutral, resilient, and just economy.

Globally, sustainable finance is closely tied to the United Nations Sustainable Development Goals (SDGs), which outline 17 ambitious targets, including poverty reduction, gender equality, and climate action. Achieving the SDGs requires an estimated US$5 to US$7 trillion annually, with US$2 to US$3 trillion of this funding expected to come from the private sector (HSBC, 2020). Financial institutions such as HSBC play a pivotal role in this transition, defining sustainable finance as financial services that integrate ESG criteria into investment decisions and actively supporting clients to reduce carbon emissions and transition to a low-carbon economy.

In Vietnam, sustainable finance research is still in its nascent stage. Key areas for further exploration include developing standardized definitions, addressing ESG risks, and creating tools to build a sustainable financial market. Additionally, there is a need to examine the role of sustainable finance in tackling climate change, enhancing biodiversity resilience, and supporting Vietnam's economic growth within the framework of sustainability.

In conclusion, sustainable finance refers to integrating ESG criteria into financial services to achieve long-term benefits for society and the environment. A clear and standardized understanding of sustainable finance is crucial for fostering research, especially in emerging markets like Vietnam, to align financial practices with global sustainability goals.

## 2.2 Activities of Sustainable Finance

Sustainable finance encompasses a wide range of activities that aim to integrate environmental, social, and governance (ESG) principles into financial systems and practices. These activities include sustainable funds, green bonds, sustainable bonds, impact investing, microfinance, active ownership, financing sustainable projects, and fostering systemic changes toward sustainability within the financial system. Below is an overview of some of the most prominent activities in sustainable finance.

### 2.2.1 Sustainable Funds

Sustainable funds are investment vehicles that apply ESG criteria to evaluate investments and assess their societal impact. According to Pan (2016), these funds aim to align financial returns with positive social and environmental outcomes. Unlike traditional funds that may only exclude specific industries (e.g., tobacco, alcohol, or gambling) based on value-based criteria, sustainable funds incorporate ESG factors into the investment process, offering additional insights that can help mitigate financial risks (HSBC, 2020). This integration demonstrates the value-driven aspect of sustainable investing while supporting long-term sustainability goals.

### 2.2.2 Green Bonds

Green bonds are fixed-income instruments designed to raise funds for climate-related and environmental projects. As defined by Liew et al., (2014), green bonds are typically asset-linked and backed by the issuer's balance sheet, carrying the same credit rating as other debts of the issuer. They fund initiatives such as energy efficiency, clean transportation, sustainable agriculture, and climate change mitigation.

Green bonds often come with tax incentives, such as exemptions or credits, making them attractive to investors while addressing emerging social challenges like climate change. Certification by third-party entities, such as the Climate Bond Standards Board, ensures that the proceeds are allocated to projects with verified environmental benefits.

### 2.2.3 Sustainable Bonds

Sustainable bonds extend the green bond framework by financing a combination of both green and social projects. These instruments are gaining traction as they align with the UN Sustainable Development Goals (SDGs) and address broader sustainability challenges (HSBC, 2020). The growth of sustainable bonds reflects the evolving financial markets' focus on social assets alongside green assets, particularly in regions like Southeast Asia, where increased flexibility supports sustainable development.

As investors demand more impact-driven investments, sustainable bonds have become a cornerstone of sustainable finance portfolios. Their development underscores the financial sector's transition toward diversified instruments such as ESG performance-linked loans and blended finance models that balance financial risk with social and environmental returns.

## 3 Overview of Studies and Research Methods in Sustainable Finance

While research on corporate social responsibility (CSR) and ethical behavior in many industries is well-established, there remains a notable gap in studies focusing on sustainable finance. This area of research is increasingly critical as sustainability becomes a significant operations management issue faced by corporations (Pan, 2016).

Sustainability has emerged as a key consideration for all sectors, driven by environmental challenges, heightened awareness of social responsibility, and the need to balance profitability. However, sustainability practices differ across industries due to their unique challenges. For instance, the electronics sector grapples with human rights, labor violations, and product recycling, while the automotive industry focuses on fuel consumption and emissions (Pan, 2016). Consequently, practices relevant to manufacturing sustainability may not directly apply to the service sector, where differences in investment rationale, institutional treatment, and performance measurements necessitate distinct approaches (Lai and Wong, 2012).

In the financial services sector, sustainability practices are still in developmental stages compared to manufacturing (Pan, 2016). This chapter highlights the financial sector's pivotal role in any economy, emphasizing its direct impact on economic growth. A stable financial system underpins sustainable development, and financial organizations contribute significantly by financing economic and developmental activities (Liew et al., 2014).

Studies on CSR in banking often explore themes like managing risk, strengthening ethics, customer protection, and establishing customer complaint mechanisms. For example, Yeung (2011) analyzed CSR reporting in the banking industry, focusing on ethics and stakeholder engagement. Similarly, Graafland and Van de Ven (2011) emphasized the need for professionalization, transparency, and strong ethical codes to enhance banks' social responsibility.

Using text mining, Pan (2016) investigated sustainability reports, identifying significant issues in Europe and North America, such as client centricity, governance, financial crime control, and sustainable mortgage and loan policies. The demand for sustainability reporting is growing, although it remains less extensive in the United States compared to Europe and Asia. The Governance & Accountability Institute reported that approximately 75% of S&P 500 and Fortune 500 firms publish sustainability reports (Pan, 2016; Baskin et al., 2014). However, annual financial performance reports alone are insufficient for informed decision-making. Financial institutions are encouraged to publish reports highlighting their long-term economic growth capacity with a focus on ethical, social, and environmental aspects.

Incorporating environmental and social challenges into financial sector activities has led to practices such as responsible lending and environmental credit risk assessments. Recently, CSR reporting has gained recognition as an effective tool for engaging stakeholders, showcasing company performance, and inspiring others (Pan, 2016; Yeung, 2011). Despite its advantages—such as aligning business with societal interests—there remains a paucity of research on key CSR components in financial services, particularly on corporate sustainability reports (Pan, 2016).

Most studies in sustainable finance leverage text mining to analyze research trends, revealing sustainability practices and trends from unstructured data. For instance, Modapothala et al (2010) analyzed corporate sustainability reports, identifying sector-specific disclosures with a focus on environmental issues. Similarly, Shahi et al (2011) employed intelligent software to score sustainability reports within the Global Reporting Initiative framework. Research by Liew et al (2014) and Pan (2016) using text mining identified common sustainability priorities in the financial sector, including health and safety, human rights, GHG reduction, energy efficiency, and community investment.

In summary, text mining has emerged as the dominant method in recent studies on sustainable finance, enabling researchers to extract and analyze key sustainability trends effectively (see Table 9.1).

## 4 Research Trend of Sustainable Finance in Vietnam

As global sustainability challenges intensify, financial institutions have a unique opportunity to drive the transition toward a more sustainable economy, adding significant value to both society and the economy. There is an increasing need for leaders capable of guiding financial institutions through the pressures brought on by these crises, as trends in sustainable finance continue to evolve. In this context, research into sustainable finance plays a pivotal role in supporting the transformation process. This section explores the emerging research trends in sustainable finance, with a focus on the transition toward a more sustainable financial market, particularly in Vietnam.

**Table 9.1** Recent research studies on sustainable finance: empirical frameworks

| Research question | Authors | Methods | Variables | Conclusions |
|---|---|---|---|---|
| What are the recent research trends of sustainability in finance? | Yeung (2011) | Text-mining method | Corporate social reports | The most prominent research trends in sustainable finance include examining the complexities of financial services, risk management, strengthening ethical practices in banking, customer protection, and establishing effective channels for addressing customer complaints |
| | Liew et al. (2014) | Text-mining method | Corporate social reports | Financial organizations play a significant role in advancing sustainable development by financing economic growth and global development initiatives |
| | Graafland and Van de Ven (2011) | Text-mining method | Company social reports | The paper suggested that emphasizing the social responsibility of banks necessitates further professional development for financial sector employees, enhanced transparency in interactions with clients and partners, the establishment of robust ethical codes, and a stronger focus on stakeholder interests |
| | Pan (2016) | Text-mining approach | Corporate sustainability reports | The key sustainability concerns in the financial sector across Europe and North America include client-centric approaches, strong governance, financial crime prevention, sustainable mortgage, and loan policies, as well as ongoing training and education |

Source: Authors' compilation

## 4.1 *Climate Change and Biodiversity Risks*

The first major trend in sustainable finance research is the growing focus on climate change and biodiversity risks. According to the 2019 World Economic Forum, extreme weather events and the failure of climate action are among the greatest global risks, with biodiversity loss following closely behind. The COVID-19 pandemic has underscored the vulnerability of industries worldwide, demonstrating that no sector is immune from disasters that can disrupt economies and wipe out years of financial gains. As the recognition grows that climate change is not just an environmental issue but a multifaceted challenge impacting all sectors, sustainable finance must address the risks associated with climate change and biodiversity loss. This trend highlights the need for continued research on the economic and financial impacts of environmental degradation and how the finance sector can respond effectively.

## *4.2 ESG (Environmental, Social, and Governance) Resilience*

The second research trend focuses on the resilience of ESG factors in finance. The shift toward low-carbon and sustainable investments is gaining momentum as a response to global environmental risks. As investors increasingly incorporate ESG criteria into their portfolios, they aim to mitigate social and environmental risks, while strengthening financial resilience. The COVID-19 pandemic has further emphasized the importance of ESG investing, as companies with strong ESG practices were found to be more resilient during the crisis. This trend calls for continued research into the impact of ESG investing on financial markets, particularly in the post-pandemic era, to better understand its role in building long-term resilience and sustainable growth.

## *4.3 Impact of Sustainable Finance on Economic Growth*

The third research trend examines the broader impact of sustainable finance on global and national economies, particularly in Vietnam. Sustainable investments help reduce financial and environmental risks while contributing to economic stability. Evidence suggests that strong ESG analysis supports resilient financial performance, making sustainable finance essential for economic growth. Moreover, investors are increasingly focused on ensuring their investments align with the United Nations Sustainable Development Goals (SDGs). This shift toward impact investing highlights the need for further research into how sustainable finance can foster economic growth and development, both globally and at the national level.

## *4.4 Tools for Sustainable Finance: Big Data, AI, and Digital Inclusion*

The fourth trend emphasizes the tools necessary for advancing sustainable finance, particularly the use of big data, artificial intelligence (AI), and digital inclusion. As the demand for social and environmental data increases, financial institutions that can harness these technologies will gain a competitive edge. Big data and AI can be utilized to track environmental trends and the performance of financial institutions, while blockchain can ensure data integrity and transparency. Furthermore, digital inclusion is becoming increasingly important for providing access to financial services, particularly for vulnerable populations. Research into the role of digital finance in promoting financial inclusion and supporting sustainable development is vital, especially in the context of overcoming the digital divide exacerbated by the COVID-19 pandemic.

### *4.5 Financial Innovation for Sustainability*

The fifth research trend is centered on financial innovation for sustainability. Following the 2008 financial crisis, which highlighted some of the self-serving aspects of the financial system, new financial instruments aimed at addressing social and environmental challenges have emerged. Green bonds, for example, have gained significant traction in the sustainable finance market, helping to finance projects with positive environmental impacts. Since the issuance of the first green bond in 2007, the market has seen significant growth. Further research into innovative financial products and strategies that support sustainability will be crucial for advancing the sustainable finance agenda in Vietnam and beyond.

### *4.6 Business Models for a Sustainable Financial Market*

The final research trend focuses on identifying the best business models for a sustainable financial market. Investors are increasingly looking for companies that demonstrate how green and sustainability bonds contribute to their transition strategies. As sustainability strategies become a key consideration for investors, there is a growing demand for greater transparency and disclosure around companies' ESG ratings and their exposure to climate risk. Research into business models that align with sustainability principles, as well as the regulatory frameworks that govern sustainable finance, will be critical for supporting the transition to a sustainable financial market.

In conclusion, sustainable finance in Vietnam is evolving rapidly, driven by both global trends and the specific needs of the Vietnamese economy. As research in these areas progresses, it will be crucial for scholars, financial institutions, and policymakers to continue exploring innovative solutions and strategies that can support a sustainable financial market, ensuring long-term economic stability and environmental stewardship.

## 5 Conclusions and Recommendations

This chapter delves into global research trends in sustainability within the finance sector. It aims to establish a clear and standardized definition of sustainable finance while exploring some of its popular applications. Additionally, the chapter reviews key studies on sustainability in financial markets and projects essential research directions, particularly in the context of the Covid-19 pandemic. A comprehensive analysis reveals that mainstream finance research has long prioritized traditional financial principles and tools, assuming that free and competitive markets inherently benefit individuals working in their self-interest. However, the persistent challenges of negative externalities and market failures—historically overlooked in finance—are now at the forefront of global concern. These include the devastating toll of the coronavirus pandemic, climate change's existential threat to the planet, widening economic inequality, and rising unemployment rates across nations.

Given these pressing issues, it is crucial for researchers and scholars to reconnect finance with its fundamental purpose. Sustainable finance offers a framework to align financial systems with global environmental and social objectives. The academic and professional spheres are increasingly embracing sustainable finance practices as demand grows for financial institutions and markets to prioritize diverse workforces, create socially responsible products, and minimize environmental harm. Concurrently, global businesses are shifting their focus toward transparency and the authenticity of their environmental and social impact claims. Scholars must carefully define sustainable finance and its core drivers to advance research in this area effectively. The authors suggest several recommendations to promote a deeper understanding and integration of sustainable finance into academic and practical domains as follows:

**Clarifying terminology in sustainable finance** Scholars and researchers should focus on precisely defining terms related to sustainable finance. Establishing clear boundaries for approaches within this field, such as differentiating "impact investing" from "environmental, social, and governance (ESG) investing," is essential. Without appreciating the nuanced distinctions, researchers risk overstating the potential of sustainable finance to address global challenges. Impact investing, as defined by Pan (2016), aims to generate measurable social and environmental benefits alongside financial returns. This approach often involves private markets, targeting early-stage ventures, nonprofits, small businesses, and public–private partnerships. While impact investing constitutes a smaller segment of sustainable finance than ESG investing, its per-transaction impact is significant. Establishing a precise and standardized definition will provide researchers with a clearer vision and enhance the quality of research in this promising field.

**Incorporating sustainable finance into academic curricula** Sustainable finance should become a core component of business education, rather than an elective topic. While leading business schools have started offering courses on impact and ESG investing, these subjects are often treated as supplementary rather than integral to the curriculum. Consequently, many finance students graduate with limited exposure to the broader societal and environmental implications of their field. Introductory finance courses could integrate topics such as calculating the social and environmental costs in net present value (NPV) analyses, valuing green bonds, and assessing systemic risks like climate change in investment portfolios. By embedding these discussions in required courses, students will be better equipped to lead initiatives that direct capital toward addressing critical global challenges.

**Linking sustainable finance with other disciplines** Advancing sustainable finance requires interdisciplinary collaboration. Research on sustainable finance is increasingly appearing in public policy, social work, law, environmental science, and other fields, enriching the dialogue, and highlighting previously ignored externalities. Expanding knowledge of sustainable finance across disciplines ensures that future leaders from various sectors can leverage financial solutions to address societal and ecological challenges. Foundations and organizations pioneering impact investing stand to benefit from a workforce with deeper expertise in sustainable finance.

**Aligning academic research with practical finance** Sustainable finance research must align more closely with corporate and investment practices. While scholars have begun integrating social and environmental goals into traditional asset pricing models, significant challenges remain. These include limited access to reliable data, particularly for impact investing, and the lack of interdisciplinary approaches in academic research. Researchers need to bridge the gap between financial metrics, such as cash flows, and broader social indicators like greenhouse gas emissions, gender equity, and economic inclusion. Notable advances have demonstrated the value of studying sustainable finance, encouraging more researchers to engage with this impactful field.

**Reframing education and research for sustainable finance** For the future of our planet, sustainability must underpin all financial activities. Achieving this transformation requires collective efforts from educators, researchers, practitioners, and policymakers. Business education programs should move beyond narrowly focusing on profit maximization to address societal imperfections and systemic risks. Future managers must understand the importance of long-term, inclusive strategies over short-term financial gains. By fostering a new generation of leaders committed to sustainable finance, we can build a more equitable and democratic financial system, driving meaningful change in response to today's global challenges.

In conclusion, integrating sustainability into finance is not merely a trend but an urgent necessity. By redefining education and research priorities, the financial sector can transition toward a model that supports societal and environmental well-being, ensuring a more sustainable future for all.

## References

Baskin D, Jordan N, Katz K (2014) The current state of sustainability reporting in financial services. Grant Thornton. Available online at http: www.granthornton.com

EU (2020) European commission's report of sustainability in finance. Available at https://finance.ec.europa.eu/sustainable-finance/overview-sustainable-finance_en

Graafland J, Van De Ven B (2011) The credit crisis and the moral responsibility of professionals in finance. J Bus Ethics 103:605–619. https://doi.org/10.1007/s10551-011-0883-0

HSBC (2020) HSBC's 2020 report of sustainable finance. Available online at https://www.hsbc.com/our-approach/esg-information

Lai K, Wong C (2012) Green logistics management and performance: some empirical evidence from Chinese manufacturing exporters. Omega 40(3):267–282. https://doi.org/10.1016/j.omega.2011.07.002

Liew WT, Adhitya A, Srinivasan R (2014) Sustainability trends in the process industries: a text mining-based analysis. Comp Indust 65:393–400. https://doi.org/10.1016/j.compind.2014.01.004

Modapothala R, Issac B, Jayamani E (2010) Appraising the corporate sustainability reports – text mining and multi-discriminatory analysis. In: Innovations in computing sciences and software engineering, vol 2010. Springer, Dordrecht, pp 489–494. https://doi.org/10.1007/978-90-481-9112-3_83

Pan Y (2016) Sustainability trend in financial services sector: evidence from Europe and North America. 2016 13th International Conference on Service Systems and Service Management (ICSSSM). https://doi.org/10.1109/ICSSSM.2016.7538549

Shahi AM, Issac B, Modapothala T (2011) Analysis of supervised text classification algorithms on corporate sustainability. International conference on computer science and network technology, Harbin. https://doi.org/10.1109/ICCSNT.2011.6181917

Yeung S (2011) The role of banks in corporate social responsibility. J Appl Econ Bus Res 1(2):103–115

**Huong-Thi Thu Le** is currently working at the Department of Political Science, University of Alberta, Canada. Her research is in international economics, applied econometrics, international trade and labor economics. The current project deals with the impact of international trade on self-employment.

**Assoc. Prof. Ly-Thi Bui** holds a PhD degree at Foreign Trade University Vietnam and she is also working as a lecturer of International Trade at Foreign Trade University. Her research interests are international trade and international economic relations.

# Part III
# Social Sustainability

# The Integration of Sustainability and CSR as a Concept in the Perception of MNCs Managers in Vietnam

**Tra-Thi Dan Vu and Phuong-Thanh Phung**

**Abstract** This chapter examines how developed countries' multinational companies interplay perceptively between corporate social responsibility (CSR) and sustainability in developing transitional economies. The chapter presents findings from a study on CSR perceptions of MNC subsidiaries in Vietnam.

The study draws on the theories of CSR pyramid and triple pillars of sustainability and employs a multi-case study design. CSR perceptions of seven well-established MNC subsidiaries in Vietnam were examined. The findings indicate that CSR perceptions covered four dimensions of the pyramid framework—economic, legal, ethical, and philanthropic—as well as sustainability-oriented responsibilities. It is notable that the four dimensions of the CSR pyramid were presented in a specific manner with an overarching responsibility toward sustainability.

Therefore, it is implied that CSR should be understood from a broad perspective that includes sustainability. The three pillars of sustainability—economic, social, and environmental—are an umbrella covering the business's economic, legal, ethical, and philanthropic responsibilities. In addition, sustainability-oriented responsibility is an overarching responsibility toward both the company's and the community's sustainability.

From the research findings, we propose a modified CSR pyramid with a new overarching dimension of sustainability responsibility. The research also adds to a growing body of literature on CSR-sustainability of MNC operating in developing countries. With the proven link between the managers' perceptions and the organization commitments and practices, it could be said that the grounds are there for actions toward the sustainability goal to take place.

---

T.-T. D. Vu (✉)
Foreign Trade University, Binh Thanh, Ho Chi Minh City, Vietnam
e-mail: vuthidantra.cs2@ftu.edu.vn

P.-T. Phung
RMIT University, Ho Chi Minh City, Vietnam
e-mail: phuong.phung@rmit.edu.vn

 
H.-T. Vu, L.-P. Dana (eds.), *Vietnam on the Road to Sustainable Economic, Social and Business Development*, Frontiers in South and Southeast Asian Development Research, https://doi.org/10.1007/978-981-96-4371-4_10

## 1 Introduction

Corporate social responsibility (CSR) has received heighted attention by both academics and practitioners in recent times (Zhu and Zhang, 2015; Carroll, 2015; Taneja et al., 2011). CSR has played an increasingly crucial role in shaping business models of many organizations due to the fact that modern-day businesses are expected to deliver beyond mere profit making and legal compliance (Carroll, 2015). Some good practices of CSR have been prevalent in developed countries, and overall CSR awareness is on the rise and demands for CSR is widespread (Hah and Freeman, 2014). Apart from its social good, CSR has been considered as a source of competitive advantage, and as an effective tool in management (Muhammad Mohtsham and Faria, 2012). Within a managerial perspective, the concept of CSR is understood as "actions that appear to further some social good, beyond the interest of the firm and that which is required by law" (McWilliams and Siegel, 2001, p. 117).

MNCs have been expected to implement CSR practices to have "license to operate" in host countries (Baskin, 2006; Jamali and Neville, 2011; Jamali et al., 2015; Young and Makhija, 2014) or required to become more socially responsible (Carroll, 2015; Carrol and Shabana, 2010; Athanasopoulou, 2012; Young and Makhija, 2014). In contrast, implementing CSR may bring about some benefits to MNCs. Research demonstrates that MNCs' socially oriented contributions can also help them build corporate image and gain host country support (Kostova et al., 2008). In addition, MNCs are increasingly becoming aware of the values of CSR and even treat CSR as a strategic instrument for gaining competitive advantage (Park and Ghauri, 2015; Porter and Kramer, 2002, 2006; Branco and Rodigues, 2006; Muhammad Mohtsham and Faria, 2012; Hart 1995). In an increasingly competitive and turbulent environment, MNCs have become more conscious of the role of CSR that is beyond mere profit making to be successful in host countries (Tixier, 2003). As a result, these stakeholders' expectations need to be integrated into policies, decisions and operations of a socially responsible business to help firms to survive in the competitive and dynamic business environment (Carroll, 2015).

Most studies in the field of CSR have focused on corporate social performance (Wood, 1991; Wartick and Cochran, 1985; Siltaoja, 2014). However, MNCs' performance in a host country is conventionally emphasized from an economic perspective, for example, how MNC maximize economic returns (Meng et al., 2014). Social adaptation (the extent that an MNC subsidiary customizes its CSR practices to respond to a host country's specific expectations) plays an important role in influencing MNCs' performance. Yet the question of how MNCs simultaneously earn profits for their shareholders while meeting social needs receives insufficient attention in both CSR and international business literature (Meng et al., 2014; Tan, 2009; Tan and Wang, 2011). In addition, while most literature has centered on the CSR of MNCs in developed countries (Yin and Zhang, 2012), the literature on it is considered to be in the infant stage (Park and Ghauri, 2015; Kolk and van Tulder, 2010). Furthermore, the perceptions and practices of social responsibility of MNC subsidiaries in developing countries have not been fully explored (Jamali, 2014; Kolk, 2016).

Furthermore, the concept of CSR "has travelled not only in meaning and application but also in cultural and institutional context" (Moon and Shen, 2010, p. 615). CSR may be interpreted differently in different national environments (Campbell, 2007; Donaldson and Dunfee, 1994). Therefore, the superimposition of the Western notions of CSR into the reality in the context of developing countries is fraught with difficulty. In other words, it is almost impossible to have a universal framework for CSR among developed and developing countries. Vietnam is a populous and market-oriented socialist country in the dynamic South East Asia and makes an interesting case to delve into the MNCs' perception of CSR practices.

Until now, CSR, a Western-born concept, remains elusive, contested (Jamali, 2007). The concept of CSR has been revolving overtime and necessary to be revisited (Carroll, 2016). Even it overlaps with other terms such as corporate citizenship, sustainability, business ethics, and stakeholder management (Carroll, 2015).

Leisinger (2007) classified CSR according to different levels of obligations and moral qualities. For example, philanthropy is at the top of the "corporate pyramid" as the "can" dimension, legal and regulatory aspects as the "must" dimension, and beyond legal minima as the "ought to" dimension. Another study examined the 37 most cited definitions of CSR via Google, of which most appear in articles and webpages of organizations such as the European Commission and CSRWire (Dahlsrud, 2008). Using content analysis, this study identified five dimensions of CSR: stakeholder, economic, social, voluntariness, and environmental dimensions. However, the link between these CSR dimensions and sustainability was not discussed.

Therefore, how developed countries' multinational companies interplay perceptively between corporate social responsibility (CSR) and sustainability in developing transitional economies is critical to be examined.

This chapter is structured as follows: (1) Section 2 provides literature on the evolution of CSR and sustainability concepts and propose a framework to examine the MNCs' perceptive interplay, (2) Sect. 3 demonstrates the research method utilized, (3) Sect. 4 presents the findings and discussions, and (4) Sect. 5 concludes.

## 2 Literature Review

### 2.1 *The Evolving CSR Concept and Its Various Dimensions*

CSR is one of the "earliest" and "key"' concepts in the literature which links the relationship between business and society (Windsor, 2001). Despite many attempts to define what CSR means, the concept of CSR is considered to "mean something, but not always the same thing to everybody" (Votaw, 1972, p. 9); the definitions of CSR in the early trajectory are rather general (Carroll, 2016). For example, CSR was defined as "the obligation of businessmen to pursue those policies, to make those decisions, or to follow those lines of actions which are desirable in terms of the objectives and values of our society" (Bowen, 1953, p. 6).

Over time, proponents of CSR have attempted to clarify "what is the right thing to do" for companies. Although consistent with Bowen (1953) on the stance of societal obligation, Davis (1960) examined the social responsibility of a business from a different viewpoint that is "social responsibility of businessmen needs to be commensurate with their social power" (Davis, 1960, p. 71). In other words, a society grants a social license and power to a business and expects in return socially responsible behaviors from it. If businesses do not fulfil their societal obligations, they will lose their legitimacy and power to operate. The social responsibilities of business in Davis's thesis (1960) include social-economic and social-human obligations. Sharing the view of Davis (1960), Frederick (1960) emphasized the fact that a business needs to use organizational resources such as financial and human resources for the betterment of the whole community, not just for the business owners. Similarly, the definition of CSR proposed by McGuire (1963) is that the responsibilities of companies are "not only economic and legal obligations but also certain responsibilities to society which extend beyond these obligations" (McGuire, 1963, p. 144). Furthermore, social responsibilities can be related to employees, education, and community welfare (McGuire, 1963). Obligations that are beyond economic and legal areas became the issue of discussions several years later. Acknowledging the previous work (Steiner, 1975) categorized the social responsibilities of businesses into two groups: internal and external. Internal social responsibilities directly concern employees and the use of resources. On the other hand, external social responsibilities focus on societal issues. Davis's later work (1973, p. 313) redefined CSR as companies' considerations and actions beyond narrow economic, technical, and legal requirements, with the claim that "social responsibility begins where the law ends." This definition emphasizes the voluntarism of companies in implementing their social responsibilities, while it does not point out what the specific actions should be.

In contrast, opponents to CSR have argued that the only responsibility of a corporation is making money. For example, "social responsibility echoes as a new tyranny of fad and fancy" and "it stops at the pocketbook" (Levitt, 1958, pp. 42–43). He concluded that "business has only two responsibilities—to obey the elementary canons of everyday face-to-face civility (honesty, good faith) and to seek material gain" (Levitt, 1958, p. 49). In other words, in this view long-term profit maximization is the critical purpose of companies. Similarly, Friedman (1962) proposed a starkly business-centric view within a free market economy. He stated that "there is one and only one social responsibility of business- to use its resources and engage in activities designed to increase its profits so long as it stays within the rules of the game, which is to say, engages in open and free competition, without deception or fraud" (Friedman, 1962, p. 133). As such, businesses can be absolved from undertaking social responsibility for their multiple stakeholders, except their shareholders. However, Friedman later posited that the responsibility of executives is "to make as much money as possible while conforming to the basic rules of society, both of those embodied in the law and those embodied in ethical custom" (Friedman, 1970, p. 32). This statement implies economic, legal, and ethical aspects of corporate responsibilities.

Acknowledging the contributions of previous researchers, Carroll was the seminal thinker in the field who in 1979 introduced a basic model, which he later refined to the pyramid, used today. Carroll (1979) attempted to determine corporate social responsibility from the perspective of four domains including economic, legal, ethical, and discretionary responsibilities. The four-dimensional construct of CSR is clearly based on considering the expectations of various stakeholders. In addition, the social expectations tend to change or evolve over time (Carroll, 2004, p. 116). In the context of globalization, it is expected that companies will become global citizens. Carroll (2004) reproduced and portrayed the CSR pyramid that still includes four dimensions of corporate social responsibility including economic, legal, ethical, and philanthropic responsibilities. Recently, Carroll (2016) revisited the construct of CSR. Notably, these four responsibilities should not be understood as being presented in a hierarchical order nor should they be attempted to be fulfilled following a hierarchical order (Carroll, 2016, p. 6; Schwartz and Carroll, 2003). With the four components of CSR, the priorities in expectations of different stakeholders are examined (Carroll, 1991, 2016). In his revision of the pyramid, Carroll (2016) considered sustainability as inseparable from CSR because there is an interchangeable usage of the two concepts in real life. Carroll (2016) also argued that the pyramid should be a "dynamic," "adaptable," and "not static" framework. Specifically, these responsibilities should be considered as an integral part of both present and future stakeholders' expectations.

Although Carroll (1979, 1991, 2004) initially suggested four main responsibilities toward a society, namely economic, legal, ethical, and philanthropic responsibilities, other scholars have offered alternative classifications of the CSR domains. Schwartz and Carroll (2003) argued that the discretionary or philanthropic domain in the pyramid does not amount to a responsibility, so it is not proper to leave it as a separate category of CSR. Instead, they suggested that the discretionary/philanthropic domain could be an add-on category to ethical responsibility. As such, they proposed an alternative approach that integrates the three core domains of economic, legal, and ethical responsibilities in a Venn model framework, resulting in seven CSR categories that overlap the three core domains. However, this model is considered to be complex and the core concepts remain intact (Melé, 2008).

## 2.2 *Sustainability: Another Evolving Concept Applied to Different Scopes*

The concept of sustainability emerged in the 1980s. Historically, the concept was derived from the idea of the World Commission on Environment and Development (WCED) of "sustainable development" in the Brundtland report (WCED, 1987).

Sustainability is defined as “development that meets the needs of the present without compromising the ability of future generations to meet their own needs” (WCED, 1987, p. 54). Sustainability emphasizes the concerns of the natural environment and the future (Carroll, 2015). The concept of sustainability has become more popular in the field of management since the notion of the “triple bottom line” (TBL) was introduced by Elkington (1998b) and linked to the idea of sustainable business (Carroll, 2015). The three pillars of TBL refers to the impacts of business practices on the economic system, their benefits to human resources and the community, and nonnegative impacts on environmental resources for future generations. Accordingly, sustainability refers to concerns for pursuing economic prosperity, environmental quality, and social justice simultaneously, sometimes called “profits, planet and people.” These concerns are found in the notion of the “triple bottom line,”, which emphasizes the long-term perspective regarding social, environmental, and economic issues.

Sustainability has been studied and applied to different scopes, from corporate or business sustainability to the geographic one such as a city, or a global community, or to the fundamental level of different types of resources, including, but not limited to, natural resources and human resources (Ramos et al., 2020). In other words, sustainable development is sustainability at a macro level (Ebner and Baumgartner, 2006) for the broader society, while corporate or business sustainability is at the micro level of that corporation or business itself.

## 2.3 *Two Chosen Prominent Frameworks: CSR Pyramid and TBL or Corporate Sustainability*

As seen from literature review, two most popular models emerged: the structural pyramid of CSR by Carroll (1991) and the Triple Bottom Line by Elkington (1998a). When Elkington coined this concept in 1998, he was starting from an accounting aspect with bottom line meaning the profit, so sustainability as he mentioned is that of a business or micro level.

The structural pyramid of CSR concept by Carroll (1991) is regarded as a prominent pattern in the CSR management field (Carrol and Shabana, 2010; Schwartz and Carroll, 2003). This model is the most well-established and widely recognized framework (Clarkson, 1995; Vissė, 2006; Wood, 1991, 2010). This model is also used in a broad range of empirical studies to explore CSR practice in developing countries (Bui, 2010; Hsu and Cheng, 2012; Jamali, 2008; Lindgreen et al., 2009). For example, the pyramid framework has been validated in previous empirical studies by Aupperle et al (1985), Edmondson and Carroll (1999), Pinkston and Carroll (1996), Burton et al (2000). These studies carried out surveys with different samples and their results gave support to the four components of the CSR pyramid. Notably, through this framework a matrix of stakeholder responsibility for social performance improvement was also discussed (Carroll, 1979, 1991). Moreover, although the pyramid of global CSR developed by Carroll (2004) attempts to adhere to the notion of the global corporate citizen, the concept of citizenship in this framework might be understood narrowly as being equivalent to CSR (Matten and Moon 2008). Within a managerial approach, Carroll's model is expected to be applicable in business–society relationship management (Meynhardt and Gomez, 2016, p. 5).

In practice, the concept of CSR has become widely supported and advanced by a multiplicity of constituents in society, even international organizations (Rosamaria and Robert 2011). Many CSR definitions are provided by governments and global entities such as the United Nations and European Commission. The definitions from these institutes are broader and toward sustainable development. For example, the UN (2007) issued a comprehensive definition of CSR at different levels:

> A minimum standard for CSR might be that businesses fulfil their legal obligations, or if laws or enforcement are lacking, that they 'do not harm'. A median approach goes beyond compliance, calling for businesses to do their best, where a 'business case' can be made, to contribute positively to sustainable development by addressing their social and environmental impacts, and potentially also through social or community investment. A maximum standard point toward their active alignment of internal business goals with externally set societal goals (those that support sustainable development).

While the Elkington 1998's TBL provides a framework for driving the business toward corporate sustainability, in a similar direction, sustainability applied at community level as reflected in the UN's 2030 agenda with 17 in its sustainable development goals (SDGs), detailed into 169 targets of people, planet, prosperity, peace, and partnership at different system levels also includes the same three pillars (economic, social, and environmental).

Interestingly, both concepts of CSR pyramid and TBL have been called for another look or rethink. While Carroll (2016) suggests other researchers will continue to explore the applicability of the Pyramid of CSR to different global, situational, and organizational contexts. Elkington (1998c) said "I hope that in another 25 years we can look back and point to this as the moment started working toward a triple helix for value creation, a genetic code for tomorrow's capitalism, spurring the regeneration of our economies, societies, and biosphere." The author also shared "All are configured around the TBL—dedicated to be not just 'best in the world', but 'best *for* the world'." In fact, we could consider his call to the concept of sustainability in his TBL as to be applied to a wider perspective of the community sustainability, as he mentioned that it should not just be the bottom line, sustainability should be about the "well-being of the people and the planet" earth we live on, which is exactly what the UN goals state.

As the two concepts refer to the role of the business in relationship with society toward sustainable development, this study utilized the combination of the pyramid of CSR and TBL as a starting point to examine CSR perceived by MNCs operating in transitional and developing host countries such as Vietnam which is poorly covered in the CSR literature.

## 3 Methods

Given that little empirical research has examined CSR perceptions in the context of developed home country MNCs operating in developing host countries in previous studies, an exploratory research design was necessary. This study has, therefore, adopted a qualitative approach with case study, abductive reasoning based on the philosophical stance of critical realism. To ensure access to information rich cases, with the aim of theoretical emphasis, the cases have been selected with differences in industry, ownership, and country of origin. From a list of 50 potential MNC subsidiaries developed by using a portfolio of companies of global compact signatory, Vietnamese CSR awardees and others known through media as well as authorities of industrial zones especially in the Southern key economic region that attracts most foreign investment in Vietnam, the researcher identified 16 companies to access. Out of the 16 companies that were contacted, gatekeepers from five companies refused to take part in the research because of their companies' confidentiality policy. Communication managers from another four companies agreed to share information about their CSR activities but did not allow access to more internal stakeholders due to their communication policy. Finally, seven MNCs were included in the study. These companies have operations nation-wide and are significant long-established foreign direct investment companies. The seven companies function in both manufacturing and services industries, operating in sectors including extraction and construction, textile, banking and finance, and fast-moving consumer goods such as dairy, flour, and beverage. These companies take pseudonyms.

Data collection was based on various forms including primary and secondary data. The primary data collected came from in-depth interviews and spontaneous direct and overt observation. The researcher's direct overt observations on company layout, bulletin boards, employee behavior, and general organizational climate are also used as another source of primary data. The secondary data was internal company publications such as documents and reports provided by participants to the researcher and gathered from publicly available publications such as information on companies' websites, press releases, industry publications, and relevant academic articles. The multiple sources of data should improve credibility and rich information for the research findings. The multiple data resources were utilized for data-cross checking to overcome the limitations of the interaction during interviews to enhance data triangulation, thus ensuring reliability (Yin, 2014) of the case study research method applied in this research. The data gathered from different resources are corroborated as evidence to avoid the potential overreliance on a certain document. Within the seven case companies, 33 internal stakeholders (including managers and executives) and 16 external stakeholders. The analytical techniques of pattern matching, and explanation building were applied during data analysis. Data analysis refers to examining, categorizing, tabulating, or otherwise recombining the evidence to address the initial proposition of a study (Yin, 2014).

## 4 Results and Discussions

Regarding CSR perceptions, the study provides a new understanding of CSR in association with TBL of sustainability (Table 1).

**Table 1** Interplayed CSR-sustainability perceptions at MNCs operating in Vietnam

| Dimensions | Economic pillar | Social pillar | Environmental pillar |
|---|---|---|---|
| Economic responsibility | – Being profitable for self-interest<br>– Creating values for other stakeholders | – Investing responsibly<br>– Providing good-quality goods and services<br>– Contributing tax | |
| Legal responsibility | | – Respecting for the law<br>– Committing to abiding the law | |
| Ethical responsibility | | – Going beyond the host country's labor law.<br>– Adopting international standard certificates<br>– Building self-regulating mechanisms with internal and external stakeholders | – Going beyond the host country's environment law<br>– Adopting international standard certificates |
| Philanthropic responsibility | Doing strategic philanthropy by developing shared value programs; supporting the community for long-term impacts | – Paying back to the community by doing charitable activities, improving living standards or quality of environment of surrounding community<br>– Committing to the community by collaborating with community in community-oriented activities; attracting other stakeholders to community-oriented activities<br>– Supporting the community for long-term impacts | – Paying back to the community by improving living standards or quality of environment of surrounding community<br>– Supporting the community for long-term impacts |
| Sustainability-oriented responsibility | – Pursuing corporate sustainability by gaining economic achievement for the business, mitigating negative effects of business activities on the environment, and being responsible for societal issues<br>– Pursuing community sustainability by gaining economic achievement for the community, protecting the environment, building social development projects<br>– Contributing to society's sustainability | | |

Source: Author's compilation

## 4.1 Economic Responsibility

Economic responsibility is a foundational responsibility in the CSR pyramid that helps companies work properly. However, economic responsibility in our case companies could be grouped into two typical perceptions: (1) profit-making and (2) socioeconomic contributions to the host country society. Firstly, profit-making was viewed by these case companies as a necessity, and an inevitable factor for them to continue their operation and business development in the future. The concept of profit making in the case companies also embraced "value creation for other stakeholders." This perception revealed that the economic responsibility was perceived beyond profit-making for the businesses themselves, and these case companies often considered the economic impacts on their stakeholders. The concept of economic responsibility was not bounded by the self-interest of case companies; rather, it was expanded to cover the economic interests of other stakeholders. Profit-making is the first bottom line for all businesses (Elkington, 1998a).

Secondly, in these case companies, economic responsibilities were perceived as socioeconomic contributions to society that could be grouped into four typical perceptions: (1) responsible investment, (2) the provision of good-quality products and services, (3) job creation, and (4) tax contribution. They were perceived as a part of their economic responsibilities that create value for stakeholders, the community and society as a whole, rather than only for their shareholders. This perception may be derived from their view of the interdependence between business and society.

The perceived economic responsibility embraces both profit-making for shareholders and economic contributions to society. The economic responsibility was perceived as the economic contributions to the company's and society's development. The two-way economic contributions were perceived to be not separate because of the intrinsic links between business and society.

This perception is not consistent with the arguments proposed by Friedman (1970), who posited that a company merely has responsibility for doing business profitably for its shareholders. In other words, economic responsibility is just limited to the company's self-interest. In contrast, responsibilities to society seem to be integrated into economic responsibility. This is consistent with the perception of the company's responsibility proposed by McGuire (1963). Moreover, the current study's evidence of the perceived economic responsibility suggests that the MNC subsidiaries create value for other stakeholders. For example, the case companies mentioned the provision of good-quality goods and services, rather than the provision of goods and services that society needs and wants (Carroll 1979). The needs or the demands for goods and services in society are diversified, even in terms of quality. Although the notion of "good quality" was not clarified by the MNC subsidiaries, their perceptions reveal their interest in creating value for their customers. In addition, their perception of economic responsibility is consistent with the finding of Vissė (2006), which covers providing investment, creating jobs and paying taxes.

The perception of economic responsibilities in this study, including profit-making, value creation, and economic contributions to society, suggests a new view that includes economic, social, and environmental aspects, rather than an emphasis on economic responsibility as a foundation as suggested by Carroll (1991, 2004).

## 4.2 *Legal Responsibility*

Legal responsibility is depicted as legal compliance (Carroll, 1979, 1991, 2004). As such, a business is expected to fulfill its economic activities within the accepted legal framework. The legal framework reflects "codified ethics" in a given society. Legal responsibility is argued to be a necessary part of CSR. These case companies perceived legal responsibility as compliance with host country legal requirements. Legal responsibilities were expressed as "respect for the law" or "commitment to abiding by the law." Furthermore, the perception of legal responsibility was also stipulated in the case companies' codes of conduct. For those case companies, legal compliance was its prerequisite to avoid any unnecessary errors that might influence the company's normal operation, avoiding conflicts of interest with employees. Further, at the case companies, the importance of the legal documents used in their daily business operations was perceived. Legal compliance was put under internal supervision or audit in these case companies. The reason for perceiving legal responsibility as an immutable component of business processes is that noncompliance would result in severe legal penalties and reputation damage. In other words, the case companies were expected to perform in a manner that was consistent with legal requirements.

This perception is consistent with the concept of legal responsibility proposed by Carroll (1979, 1991, 2004). Legal responsibility is reflected through respect for the law or commitment to abide by relevant laws. The laws are also clarified at different regulatory levels including international, national, and local authorities' regulations. The perception seems to emphasize the legal implementation aspect. This perception is necessary for MNC subsidiaries to avoid unnecessary legal violations during their business operations in the host country. This perception may be derived from the institutional complexity of the MNC subsidiaries' operating environment and avoiding costs caused by institutional unfamiliarity (Yang and Rivers 2009; Gifford et al. 2010). Although legal compliance implies a reactive nature, these case companies' perception of legal responsibility seems to be proactive in terms of legal implementation. This finding has a significant implication for companies that are operating and planning to invest in foreign host countries, especially in transitional economies where legal systems have been changing and gradually becoming complete.

### 4.3 Ethical Responsibility

Ethical responsibility normally embraces doing right, just and fair things that are not codified in laws Carroll (1979, 1991, 2004). These behaviors are often expected in society to avoid harms and to prevent harm to stakeholders. Such responsibilities are identified based on the societal norms and expectations of each society. Unlike legal responsibility that is codified in law, it is quite difficult to identify the boundaries of ethical responsibility. The findings highlight that the ethical responsibility embraced going beyond regulatory compliance, and focuses employee well-being, working environments, and environmental protection; adopting international standard certificates; and building self-regulating mechanisms.

Specifically, ethical responsibility was perceived as going beyond regulatory compliance to doing the right things; for example, the reduction in carbon emissions and the provision of well-being for employees were better than legally regulated. This perception is consistent with the perception proposed by Carroll (1979, 1991, 2004).

However, the notion of "going beyond regulatory compliance" seems abstract. In the current study, the case MNC subsidiaries interpreted their ethical responsibility in terms of more detailed sub-issues, that is, adopting international standards and building self-regulating mechanisms. International standards may be perceived as going "beyond national laws" (van Der Ploeg and Vanclay, 2017). As such, the MNC subsidiaries perceived the adoption of international standards as a means to help them survive and compete in an increasingly competitive business environment as well as meeting their customers' expectations (Giuliani, 2016).

Building self-regulating mechanisms was also perceived as a voluntary means to manage business operations and stakeholder relationships. These mechanisms were reflected through their codes of conduct, which presented ethical principles such as integrity, fairness, and transparency. This self-regulation reveals what are the right things to do in their operations. The finding is consistent with the findings of Vissȇ (2006). These activities seem to overcome the insufficient state regulation in the host country. Consequently, ethical responsibility helps these businesses mitigate negative impacts and unnecessary crises that state legal frameworks cannot prevent.

With the specific perceptions of ethical responsibility, it seems that the case companies perceived ethical responsibility in a proactive manner. This perception overcomes the blurry definition of ethical responsibility proposed by Carroll (1994) that is still vaguely addressed (Jamali, 2007). Moreover, in practice ethical responsibility is blurring for MNC subsidiaries operating in an unfamiliar host-country environment. These findings suggest practical implications for companies in relation to management, such as developing codes of conduct and gaining international standards.

## 4.4 Philanthropic Responsibility

Another important finding is that the philanthropic responsibility was not perceived as being limited to charitable activities; it also embraces sustainability commitment to the community and strategic philanthropy. Specifically, the study has found "paying back to the community" as one of the MNC subsidiaries' perceived community-oriented responsibilities. However, this research has also found that the case companies perceived their commitments to long-term, rather than short-term, charitable activities by expressing the purpose of their philanthropic activities as to improve the socioeconomic situation and well-being of the host-country community. While manufacturing- and mining-related case companies tended to focus on philanthropic activities in surrounding communities, service-related case companies tended to pay attention to the community at large without caring about the proximity of communities in the host-country environment. This finding of a possible difference in geographical proximity in the case companies' CSR perception may be derived from the proximity of their business activities to natural resources, and work health and safety issues. This finding is consistent with the findings of previous authors (van Der Ploeg and Vanclay, 2017; Wright, 2008).

The study has also found that the case MNC subsidiaries perceived their commitments to the community in terms of a win–win approach. As such, perceiving themselves as good citizens, the MNC subsidiaries were found to be partners with other stakeholders in contributing to the development of communities. This seems to go beyond the perception of philanthropic responsibility as an "add-on" option as proposed by Schwartz and Carroll (2003). Instead, this community-oriented commitment tended to be an integral part of their philanthropic responsibility as well as their managerial approach to networking around CSR activities. This finding further supports the ideas of the scope of CSR in networked MNCs proposed by Enderwick (2018) and social sector partnership proposed by Jayaramna et al (2018) for social development. This perception suggests the active role of the MNC subsidiaries in the host-country environment. This finding is significant in implying that MNC subsidiaries should consider the interests and concerns of different stakeholders and follow a "balanced approach" through partnership to create sustainable development for both companies and communities (Porter, 2008).

In contrast to the perception of "paying back to the community" and "commitment to the community" that seem to be perceived as voluntary responsibilities for community benefit, this research has found that strategic philanthropy was perceived to bring benefits for both the company and the community. The benefits for the company were perceived in both indirect and direct impacts, such as improvement of branding image or reputation, and reduction in costs of production, respectively. In sum, strategic philanthropy involves moving beyond charitable activities to reconcile social and economic benefits for the company. Depending on the attributes, skills, and competence of each company, social activities are selected to address social needs and self-interest such as company reputation or market opportunities (Porter and Kramer, 2011; Polonsky and Jevons, 2009). Strategic philanthropy is, thus, quite a flexible responsibility in pursuing business objectives.

While the first bottom line is closely associated with the economic dimension of the CSR concept, the other three dimensions including legal, ethical, and philanthropic aspects could be associated closely with the social bottom line. The social bottom line, specifically, is about "social capital," which in a broader sense, means the trust of the society on the company (Elkington, 1998a, b). In fact, with the research findings, when a company accomplishes its legal, ethical, and philanthropic responsibilities, the society would look at it with trust, and therefore, the company can obtain the social capital to meet the target for its social bottom line. Interestingly, the findings indicate that the environmental bottom line is associated with the ethical and philanthropic responsibilities of the corporation. Arguably, the long-term maintenance of "natural capital" could be seen as contributing to long-term economic achievements of companies, but that has not been in the perception of the managers yet.

## 4.5 Sustainability-oriented Responsibility

It is interesting to note that, in all seven cases examined in this study, the four core responsibilities—economic, legal, ethical, and philanthropic—were supported by an overarching sustainability-oriented responsibility in terms of triple economic, social, and environmental concerns. Notably, the notion of sustainability-oriented responsibility embraces both corporate sustainability and community sustainability, reflecting harmony between the benefits to the corporation and to the community. This perception seems to embrace the idea that business and society are intertwined (Frederick, 1994) and the recent trends related to sustainability across the world. In practice, sustainability trends have led to great concerns of both business and society since the UN launched the Millennium Development Goals. Before that, the concept of sustainable development was proposed in the Brundtland Report (1987) by the WCED (1987). As such, three components—environmental protection, economic growth, and social equity—are highlighted. For businesses, corporate sustainability seems to be derived from the notion of sustainable development. In other words, the economic, social, and environmental pillars seem to be perceived as responsibilities that companies aim to achieve. This finding corroborates the idea of Romero and Lamadrid (2014, p. 180), who suggested that CSR should be explained "as corporate sustainability and responsibility." This finding implies that sustainability and responsibility are complementary (Visser, 2010) and suggests a comprehensive management approach for the sustainability of both corporations and communities.

Notably, while the four components of the CSR pyramid (Carroll, 1991) still exist, the current study has found that CSR as perceived by the case companies adhered to a sustainability orientation. In other words, the CSR pyramid proposed by Carroll (1991) has been modified in association with this sustainability-oriented responsibility. This seems to reflect harmony between the MNC subsidiaries' CSR perception in the context that sustainability is discussed globally and has become an emerging concern in developing countries.

### *4.6 Limitations and Recommendations*

This research identified how MNCs interpret their CSR in a host developing country such as Vietnam. To answer this question, a comprehensive review of the extant literature on relevant debates around the CSR concept in its early trajectory, unified frameworks of CSR and the necessity of revisiting a dominant framework of the CSR pyramid in the context of developing countries provided deeper understanding of the current dimensions of CSR perceptions. However, recently the concept of sustainability has emerged. While CSR perceptions at the case MNCs tend toward longer-term social, economic, and environmental impacts, the relation between the CSR and sustainability has not been examined thoroughly, suggesting future research directions.

## 5 Conclusion

The objective of this research was to identify the perceptions of CSR at companies, specifically at MNC subsidiaries operating in Vietnam. In answering this question, a comprehensive review of the extant literature on relevant debates around the CSR concept in its early trajectory, unified frameworks of CSR, and the necessity of revisiting a dominant framework of the CSR pyramid in the context of developing countries provided deeper understanding of the current dimensions of CSR perceptions. This study utilized the CSR pyramid and TBL as a starting point to explore the perceptions of MNCs operating in Vietnam.

The findings indicated that CSR perceptions covered four dimensions of the pyramid framework—economic, legal, ethical, and philanthropic—as well as sustainability-oriented responsibilities. It is notable that the four dimensions of the CSR pyramid were presented in a specific manner with an overarching responsibility toward sustainability. For example, economic responsibility was perceived as both profit-making and economic contributions to society. Legal responsibility was understood as legal compliance with international, national, and local authorities' regulations.

In contrast, ethical responsibility was referred to as going beyond regulatory compliance, adopting international standard certificates, and building self-regulating mechanisms. Paying back to the community, committing to the community, and doing strategic philanthropy were perceived as philanthropic responsibility. Notably, sustainability-oriented responsibility focused on pursuing corporate sustainability and community sustainability. The findings suggest that, in general, the CSR pyramid proposed by Carroll (1991) can be modified by incorporating sustainability-oriented responsibility. The evidence from this research suggests that the perceptions of CSR at MNC subsidiaries tend to harmonize with emerging concerns about sustainability in both global and developing country contexts. Research question 1 relating to the first objective has therefore been answered.

The empirical findings in this research provide a new understanding of CSR in the context of sustainability being discussed globally as an emerging concern in developing countries. MNCs operating in developing countries are increasingly expected to conduct their businesses in a socially responsible manner. MNCs are under pressure to implement CSR practices in their host countries to ensure long-term business survival and competitiveness. CSR, as a Western concept, reflects a complicated phenomenon that is contextual and multimodal, and created through interaction between the company and external and internal stakeholders. This study employs a qualitative approach, case study design with seven long-established MNCs across a variety of sectors, involving 49 in-depth interviews including managers and executives who are in charge of CSR and directly involved in the companies' day-to-day CSR practices and external stakeholders. The findings reveal that CSR perceptions at these MNCs are holistic with strategic and developmental orientations, not purely philanthropic. The findings have significant implications for theory and practice of CSR, and also for the transfer of international CSR-related business practices. Findings also suggest that future research needs to be directed at exploring the connection between CSR and the emerging concept of sustainability; the influence of institutional complexity on the CSR practices of MNC subsidiaries operating in developing countries.

**Acknowledgments** This research is a part of the research project supported through funding from Swinburne University and the Vietnam's Ministry of Education and Training. The Recipient of the funding is Tra Thi Dan Vu. We would like to thank Dr. Chandana Hewege and Professor Bernadine Van Gramberg at Swinburne University of Technology for their valuable comments and suggestions.

## References

Athanasopoulou A (2012) Managers' corporate social responsibility perceptions and attitudes across different organizational contexts within the non-profit–for-profit organizational continuum. J Chang Manag 12:467–494

Aupperle K, Carroll A, Hatfield J (1985) An empirical examination of the relationship between corporate social responsibility and profitability. Acad Manag J 28:446

Baskin J (2006) Corporate responsibility in emerging markets. J Corporate Citizen 24:29–47

Bowen HR (1953) Social responsibilities of the businessman. Harper & Row, New York

Branco MC, Rodigues L (2006) Corporate social responsibility and resource-based perspectives. J Bus Ethics 69:111–132

Bui TLH (2010) The Vietnamese consumer perception on corporate social responsibility. J Int Bus Res 9:75

Burton B, Farh J-L, Hegarty W (2000) A Cross-cultural comparison of corporate social responsibility orientation: Hong Kong vs. United States Students. Teaching Business Ethics 4:151–167
Campbell J (2007) Why would corporations behave in socially responsible ways? An institutional theory of corporate social responsibility. Acad Manag Rev 32:946–967
Carrol AB, Shabana KM (2010) The business case for CSR: a review of concepts, research and practice. Int J Manag Rev 12:85–105
Carroll AB (1979) A three-dimensional conceptual model of corporate performance. Acad Manag Rev 4:497–505
Carroll AB (1991) The pyramid of corporate social responsibility: toward the moral management of organizational stakeholders. Bus Horiz 34:39–48
Carroll AB (1994) Social issues in management research experts' views, analysis, and commentary. Bus Soc 33:5–29
Carroll AB (2004) Managing ethically with global stakeholders: a present and future challenge. Acad Manag Exec 18:114–120
Carroll AB (2015) Corporate social responsibility: the centerpiece of competing and complementary frameworks. Organ Dyn 44:87–96
Carroll AB (2016) Carroll's pyramid of CSR: taking another look. Int J Corp Soc Responsib 1:1–8
Clarkson MBE (1995) A stakeholder framework for analyzing and evaluating corporate social performance. Acad Manag Rev 20:92–117
Dahlsrud A (2008) How corporate social responsibility is defined: an analysis of 37 definitions. Corp Soc Responsib Environ Manag 15:1–13
Davis K (1960) Can business afford to ignore social responsibilities? Calif Manag Rev 2
Davis K (1973) The case for and against business assumption of social responsibilities. Acad Manag J 16:312–322
van Der Ploeg L, Vanclay F (2017) A tool for improving the management of social and human rights risks at project sites: the human rights sphere. J Clean Prod 142:4072–4084
Donaldson T, Dunfee TW (1994) Toward a unified conception of business ethics-integrative social contracts theory. Acad Manag Rev 19:252–284
Ebner D, Baumgartner RJ (2006) The relationship between sustainable development and corporate social responsibility. Corporate responsibility research conference. Queens University, Belfast Dublin
Edmondson V, Carroll A (1999) Giving back: an examination of the philanthropic motivations, orientations and activities of large black-owned businesses. J Bus Ethics 19:171–179
Elkington J (1998a) Accounting for the triple bottom line. Meas Bus Excell 2(3):18–22
Elkington J (1998b) The 'Triple Bottom Line' for 21st century business. In: Starkey, Welford (eds) The Earthscan Reader in Business & Sustainable Development (2001). Earthscan Publishing, London
Elkington J (1998c) Cannibals with forks: the triple bottom line of 21st century business, Gabriola Island: John Wiley & Sons, Ltd
Enderwick P (2018) The scope of corporate social responsibility in networked multinational enterprises. Int Bus Rev 27:410–417
Frederick WC (1960) The growing concern over business responsibility. Calif Manag Rev 2:54–61
Frederick W (1994) From CSR1 to CSR2: the maturing of business-and-society thought. Bus Soc 33:150–164
Friedman M (1962) Capitalism and freedom, Chicago. University of Chicago Press, Chicago
Friedman M (1970) The social responsibility of business is to increase its profits. New York Times Magazine 13:32–33
Gifford B, Kestler A, Anand S (2010) Building local legitimacy into corporate social responsibility: gold mining firms in developing nations. J World Bus 45:304–311
Giuliani E (2016) Human rights and corporate social responsibility in developing countries' industrial clusters. J Bus Ethics 133:39–54
Hah K, Freeman S (2014) Multinational enterprise subsidiaries and their CSR: a conceptual framework of the management of CSR in smaller emerging economies. J Bus Ethics 122:125–136

Hart SL (1995) A natural-resource-based view of the firm. Acad Manag Rev 20:986–1014
Hsu J-L, Cheng M-C (2012) What prompts small and medium enterprises to engage in corporate social responsibility? A study from Taiwan. Corporate Soc Respons Environ Manag 19:288–305
Jamali D (2007) The case for strategic corporate social responsibility in developing countries. Bus Soc Rev 112:1–27
Jamali D (2008) A stakeholder approach to corporate social responsibility: a fresh perspective into theory and practice. J Bus Ethics 82:213–231
Jamali D (2014) CSR in developing countries through an institutional lens. *Corporate social responsibility and sustainability: Emerging trends in developing economies*
Jamali D, Neville B (2011) Convergence versus divergence of CSR in developing countries: an embedded multi-layered institutional lens. J Bus Ethics 102:599–621
Jamali D, Lund-Thomsen P, Jeppesen S (2015) SMEs and CSR in developing countries. Bus Soc:1–12
Jayaramna A, D'souza V, Ghoshal T (2018) NGO-business collaboration following the Indian CSR Bill 2013: trust-building collaborative social sector partnerships. Dev Pract 28:831–841
Kolk A (2016) The social responsibility of international business: from ethics and the environment to CSR and sustainable development. J World Bus 51:23–34
Kolk A, van Tulder R (2010) International business, corporate social responsibility and sustainable development. Int Bus Rev 19:119–125
Kostova T, Roth K, Dacin M (2008) Institutional theory in the study of multinational corporations: a critique and new directions. Acad Manag Rev 33:994–1006
Leisinger KM (2007) Corporate philanthropy: the "top of the pyramid". Bus Soc Rev 112:315–342
Levitt T (1958) The dangers of social responsibility. Harv Bus Rev 36:41–50
Lindgreen A, Swaen V, Campbell T (2009) Corporate social responsibility practices in developing and transitional countries: Botswana and Malawi. J Bus Ethics 90:429–440
Matten D, Moon J (2008) "Implicit" and "explicit" CSR: a conceptual framework for a comparative understanding of corporate social responsibility. Acad Manag Rev 33:404
McGuire JW (1963) Business and society. McGraw-Hill
McWilliams A, Siegel D (2001) Corporate social responsibility: a theory of the firm perspective. Acad Manag Rev 26:117–127
Melé D (2008) Corporate social responsibility theories. The Oxford handbook of corporate social responsibility. Oxford University Press
Meng Z, Park S-H, Zhou N (2014) MNC strategy and social adaptation in emerging markets. J Int Bus Stud 45:842–861
Meynhardt T, Gomez P (2016) Building blocks for alternative four-dimensional pyramids of corporate social responsibilities. Bus Soc 58(2). https://doi.org/10.1177/0007650316650444
Moon J, Shen X (2010) CSR in China research: salience, focus and nature. J Bus Ethics 94:613–629
Muhammad Mohtsham S, Faria A (2012) Corporate social responsibility as a source of competitive advantage: the mediating role of social capital and reputational capital. J Database Mark Cust Strategy Manag 19:219–232
Park BI, Ghauri PN (2015) Determinants influencing CSR practices in small and medium sized MNE subsidiaries: a stakeholder perspective. J World Bus 50:192–204
Pinkston T, Carroll A (1996) A retrospective examination of CSR orientations: have they changed? J Bus Ethics 15:199–206
Polonsky M, Jevons C (2009) Global branding and strategic CSR: an overview of three types of complexity. Int Mark Rev 26:327–347
Porter TB (2008) Managerial applications of corporate social responsibility and systems thinking for achieving sustainability outcomes. Syst Res Behav Sci 25:397–411
Porter ME, Kramer MR (2002) The competitive advantage of corporate philanthropy. Harv Bus Rev 80:56–68
Porter M, Kramer M (2006) Strategy and society: the link between competitive advantage and corporate responsibility. Harv Bus Rev 84:78–92

Porter ME, Kramer MR (2011) Creating shared value. Harv Bus Rev 89:62–77
Ramos TB, Caeiro S, Disterheft A, Mascarenhas A, Deutz P, Spangenberg JH, Montano M, Olayide O, Sohal A (2020) Rethinking sustainability: questioning old perspectives and developing new ones. J Clean Prod 258:120769
Romero KCC, Lamadrid RL (2014) Rethinking corporate social responsibility within the sustainability agenda. J Global Respons 5:180–202
Rosamaria CM-L, Robert CP (2011) Historical background of corporate social responsibility. Soc Res J 7:528–539
Schwartz MS, Carroll AB (2003) Corporate social responsibility: a three-domain approach. Bus Ethics Q 32:503–530
Siltaoja ME (2014) Revising the corporate social performance model – towards knowledge creation for sustainable development. Bus Strateg Environ 23:289–302
Steiner GA (1975) Business and society. Random House Inc.
Tan J (2009) Multinational corporations and social responsibility in emerging markets: opportunities and challenges for research and practice. J Bus Ethics 86:151–153
Tan J, Wang L (2011) MNC strategic responses to ethical pressure: an institutional logic perspective. J Bus Ethics 98:373–390
Taneja S, Taneja P, Gupta R (2011) Researches in corporate social responsibility: a review of shifting focus, paradigms, and methodologies. J Bus Ethics 101:343–364
Tixier M (2003) Soft vs. hard approach in communicating on corporate social responsibility. Thunderbird Int Bus Rev 45:71–91
United Nations (2007) CSR and developing countries: what scope for government action? Sustainable development innovation briefs. Policy integration and analysis branch of the division for sustainable development [Online]. Available from https://sustainabledevelopment.un.org/content/documents/no1.pdf [Accessed 8 November 2017]
Vissė W (2006) Re-visiting Carroll's CSR pyramid: an African perspective. In: Rahbek MHEP (ed) Corporate citizenship in developing countries-new partnership perspectives. Copenhagen Business School Press, Copenhagen
Visser W (2010) CSR 2.0: the evolution and revolution of corporates official responsibility. Responsible business: How to manage a CSR strategy successfully, 311
Votaw D (1972) Genius becomes rare: a comment on the doctrine of social responsibility Pt. I. Calif Manag Rev 15:25–31
Wartick S, Cochran P (1985) The evolution of the corporate social performance model. Acad Manag Rev 10:758–769
WCED (1987) Our common future (The Brundtland Report) [Online]. Oxford: World Commission on Environment and Development. Available http://www.un-documents.net/our-common-future.pdf [Accessed 8 November 2018]
Windsor D (2001) The future of corporate social responsibility. Int J Organ Anal 9:225–256
Wood D (1991) Corporate social performance revisited. Acad Manag Rev 16:691–718
Wood DJ (2010) Measuring corporate social performance: a review. Int J Manag Rev 12:50–84
Wright M (2008) A survey of the scope and patterns of alleged corporate-related human rights abuse. Corporate social responsibility initiative working paper. John F. Kennedy School of Government, Haward University, Cambirdge, MA
Yang X, Rivers C (2009) Antecedents of CSR practices in MNCs' subsidiaries: a stakeholder and institutional perspective. J Bus Ethics 86:155–169
Yin RK (2014) Case study research: design and methods. Directed by Yin RK
Yin J, Zhang Y (2012) Institutional dynamics and corporate social responsibility (CSR) in an emerging country context: evidence from China. J Bus Ethics 111:301–316
Young S, Makhija M (2014) Firms' corporate social responsibility behavior: an integration of institutional and profit maximization approaches. J Int Bus Stud 45:670–698
Zhu Q, Zhang Q (2015) Evaluating practices and drivers of corporate social responsibility: the Chinese context. J Clean Prod 100:315–324

**Tra-Thi Dan Vu** is a lecturer at Foreign Trade University (FTU), Ho Chi Minh City campus (Vietnam). She is a member of the research program "Environmental, social and governance activities: empirical studies on enterprises" funded by FTU. She holds a PhD in Management (International Business) from the Swinburne University of Technology (Australia). She teaches, conducts research, and consults in the areas of international trade, human resource management, corporate social responsibility, and sustainability. She has published several articles and one book chapter. She has also supervised both undergraduate and master students doing research on sustainable consumption, human resource management and international trade.

**Phuong-Thanh Phung** has been teaching at RMIT Vietnam since 2007. She works in the Department of Business Innovation, RMIT Vietnam, 702 Nguyen Van Linh, Tan Phong, District 7, Ho Chi Minh City, Vietnam. She graduated with an MBA degree from University of Southern California, the United States in 2006. Prior to that, she worked at Vistar Ltd. (Hanoi, Vietnam), Hanoi Foreign Trade University (Vietnam), and Mattel in El Segundo, California (USA). Phuong has published a few articles and contributed to one book chapter. She is pursuing the DBA program with Heriot-Watt University (UK). Her research interests include social marketing, sustainable consumption and tourism, and management. She leads as a course coordinator in industry client projects and live case studies for students, including those for the specialist course Marketing Intelligence and the capstone course Digital Marketing Strategy and Planning in the Digital Marketing Program at RMIT Vietnam.

# Environmental Corporate Social Responsibility: A Case Study in Vietnam

**Minh-Duc Le and Hien-Thi Tran**

**Abstract** Activities of business enterprises are getting more influence on the society, holding them accountable in terms of corporate social responsibility (CSR). From analyzing the case of Hung Nghiep Formosa Ha Tinh Steel Company Limited, this chapter is a case of the environmental responsibilities of business enterprises and their impacts to stakeholders, that is, environmental CSR. The chapter are presented in the four main parts: (i) understanding the CSR concept and its importance, (ii) briefly review the guidance of the International Organization for Standardization ISO 26000: 2010 on environmental CSR, (iii) figuring out which CSR issues this typical case incurs, and (iv) making recommendations for protecting environment under the umbrella of CSR.

## 1 Introduction

### *1.1 Background Context*

Over the last decades, Vietnam has gone through a period of rapid socioeconomic development. Especially, becoming a WTO member in 2007 (WTO, 2007) has offered a large number of business opportunities for the country and business enterprises. Over that period, the country has steadily and rapidly achieved annual GDP growth of 2.78%/year in 1986 and 5%/year in 2023 (World Bank, 2023). Since Vietnam's accession to the WTO, a huge foreign direct investment (FDI) has been

M.-D. Le
British University Vietnam, Ecopark, Ha Noi, Vietnam
e-mail: minh.ld@buv.edu.vn

H.-T. Tran (✉)
University of Economics and Business, Vietnam National University, Ha Noi, Cau Giay, Ha Noi, Vietnam
e-mail: hientt.hsb@vnu.edu.vn



H.-T. Vu, L.-P. Dana (eds.), *Vietnam on the Road to Sustainable Economic, Social and Business Development*, Frontiers in South and Southeast Asian Development Research, https://doi.org/10.1007/978-981-96-4371-4_11

flowing in the country in many forms, especially setting up commercial presence such as economic organizations, joint ventures, branches, or representative offices.

Nevertheless, despite the abovementioned benefits, the country is still facing environmental challenges. Specifically, urbanization, industrialization, and intensive production have been creating adverse negative impacts on Vietnam's environment. Climate change, air pollution, water pollution, noise pollution, environmental degradation, deforestation, and animal extinction will be the consequences, if these factors are not under control. As one of the main contributors to the country's sustainable development, the corporate sector has been constantly called out to take actions to respond to this situation by applying corporate social responsibility (CSR).

At international level, there are various guidelines regulating the issue, such as (i) OECD Guidelines for Multinational Enterprises, (ii) ISO 26000 Guidance on Social Responsibility, (iii) UN Global Compact, and (iv) UN Sustainable Development Goals which are voluntary and not regulatory enforcement. However, these documents only reflect the expectations and the desires at international level for enterprises in applying CSR. It is important and necessary for countries to lean on these guidelines to establish their own legal framework with respect to CSR. At the domestic level, Vietnam has been touching on this topic in several regulated documents from laws to circulars. In addition, Vietnam has actively participated and ratified several international conventions in terms of environmental protection such as the Convention on Climate (1994), the Kyoto Protocol (2002), and the Nagoya Protocol (2014).

However, it is evident that several business enterprises focus on making profit for themselves without paying attention to the benefits of other stakeholders. Consequently, serious environmental problems have been caused and brought severe impacts to all stakeholders, especially local people. A recent case of environmental pollution caused by business enterprises in Vietnam is Hung Nghiep Formosa Ha Tinh Steel Company Limited. This steel factory did not fulfill its commitment to the environment, acted irresponsibly and caused serious environmental incidents, resulting in severe impacts for people in four provinces starting from Vung Ang–Ha Tinh Province, then spread to Quang Binh province, Quang Tri province, and Thua Thien-Hue province ("Formosa takes responsibility for fish deaths," 2016).

## 1.2 Aim and Objectives of the Study

With an aim to suggest policy reform to enhance business sustainability and environmental protection in Vietnam, this chapter analyzes the case of Hung Nghiep Formosa Ha Tinh Steel Company Limited. The chapter contains four main parts including (i) understanding the CSR concept and its importance, (ii) briefly review the guidance of the International Organization for Standardization ISO 26000: 2010 on environmental CSR, (iii) figuring out which CSR issues this typical case incurs, and (iv) making recommendations for protecting environment under the umbrella of CSR.

# 2 CSR Concept and Importance

## 2.1 The Concept

The CSR conceptualization has been developed, formalized, and redefined by many authors. McGuire (1963) denoted that "The idea of social responsibilities supposes that the corporation has not only economic and legal obligations but also certain responsibilities to society which extend beyond these obligations." This notion of CSR is likely to be appropriate as it is extended beyond economic and legal obligations, luring reader attention to other responsibilities that a corporation may hold during its business operation, although this notion of CSR is not able to clarify what other responsibilities are.

Steiner (1971) created a new view on CSR by stating "Business is and must remain fundamentally an economic institution, but [...] it does have responsibilities to help society achieve its basic goals and does, therefore, have social responsibilities. The larger a company becomes, the greater are these responsibilities." From Steiner's (1971) viewpoint, it is necessary for the larger business enterprises to have greater responsibilities. The responsibilities of the enterprises should also depend on the coverage of their business lines and the magnitude of their impacts on the stakeholders.

Elaborating on the idea of social responsibility in business, Johnson (1971) pointed out that a firm can only be considered socially responsible, if its management pursuits not only the profit but also the interests of other stakeholders. Johnson (1971) also emphasized that social activities "add profits to their organization".

The year 1991 was the milestone when the concept of CSR was much more clarified. Specifically, Carroll (1991) provided a certain scope of CSR that encompasses four types of social responsibilities including the social, legal, ethical, and philanthropic expectations at a given time. This is an approach of CSR that several authors have supported and used to elaborate the CSR concept or add value to its clarifications.

The European Commission (2001) defined CSR as "a concept whereby companies integrate social and environmental concerns in their business operations and in their interaction with their stakeholders on a voluntary basis."

Matten and Moon (2004) argued that it is problematic to define CSR since it is a dynamic and always challenging concept in each particular economic, political, social context. It is notable that the concept of CSR has been developed over years; however, neither the global agreement nor legal document has been codified to provide a consistent definition to CSR (Smith, 2011).

From a shareholder's perspective, there were arguments against the social responsibilities that business enterprises should have apart from making profit. One of the most highlighted statements was "The social responsibility of business is to increase its profits" made by Friedman (1970). It means as long as business enterprises achieve this goal, they are considered "responsibly social corporate." Specifically, Friedman (1970) claimed that "Only people can have responsibilities";

business enterprises are legal entities which are set up by nature persons, therefore they may not have "artificial responsibilities," or social responsibilities which only nature persons have. Friedman's argument (1970) is controversial with most critics by the scholars with stakeholder perspective such as Carroll (1999).

From the chapter authors' perspective, Friedman's statement (1970) would have been reasonable in the way it is applicable to solely moral responsibilities as business enterprises are not moral persons. Nevertheless, given that other obligations may be expected by society, the authors view that Friedman (1970) failed to recognize the certain obligations which include legal obligations to obey the law. The law is built on the basis of public interests, therefore, if business enterprises do not comply with the law, they are very much likely to create problems for the public. This has been evident by several corporate-level violation cases in which the firms did not comply with the law or abused the law's gray areas to maximize their profits, leading to environmental disasters. As a consequence, the pollution costs and damages to the environment may be beyond their profits (Jowit, 2010). Making profits illegally is certainly an irresponsible and unacceptable action to society.

## 2.2 *Why Should Business Enterprises Implement CSR?*

In the chapter authors' view, there are two main drivers to business enterprises' decision of CSR engagement, derived from existing literature and current regulations at both international and domestic levels.

### 2.2.1 Market and Social Drivers

Business activities involve many different markets including the supply market, the labor market, the product market, and the investment market. In these markets, there are several factors/stakeholders involved in and having impacts, namely suppliers, staff, customers, and investors/shareholders. To maintain the business, business enterprises need to take into account all of not only customers' attitudes and feelings toward their service/product but also their strategy/vision and contribution to society. For instance, Starbucks was boycotted due to tax evasion in 2013. Consequently, the company's subsidiary in the United Kingdom suffered its first sales fall from 413 million pounds in 2012 to 399 million pounds in 2012, despite its 16 years' consecutive growth (Bergin, 2014). This is a clear example proving that if business enterprises fail to meet societal expectations, they are likely not to receive the society's support. This view has been shared by various authors such as Wood (1991) by stating "if central stakeholders lose confidence in the firm's performance, legitimacy may be withdrawn." The market is the place where media and competitors always exist to assess company profile and evaluate every step a business enterprise takes. Hence, pressure from these stakeholders can be considered one of the main drivers leading to CSR engagement.

### 2.2.2 Governmental Drivers

Although, CSR has not been ratified in any legal document as a legal requirement, in fact, in Vietnam and some countries in the world, requirements to implement certain activities, responsible to society and nature, have been codified into law.

With an aim to help measure, monitor, and manage performance undertaken by business enterprises and their impact on society, the Directive 2014/95/EU, Art 3 requires that large public-interest entities having an average number of employees in excess of 500 to disclose in their management report on their policies, main risks, and outcomes relating to at least "environmental matters, social and employee-related matters, respect for human rights, anti-corruption and bribery matters" (Art 5). The reports are also required to be made available to the public and authorities (Art 6).

Particularly, in accordance with the UK Companies Act 2006, directors are regulated to act in consideration of not only the success of the company, but also the interests of their employees, the community, and the environment.

In India, business enterprises "having net worth of rupees five hundred crore or more, or turnover of rupees one thousand crore or more or a net profit of rupees five crore or more during any financial year" must establish a Corporate Social Responsibility Committee that is responsible for drafting and monitoring the Corporate Social Responsibility Policy of the company from time to time (Companies Act, 2013).

In Vietnam, Circular 96/2020/TT-BTC, issued by the Ministry of Finance on November 16, 2020, provides guidelines for information disclosure in the securities market. It specifically requires public companies to prepare and disclose their annual reports within 20 days of publishing their audited annual financial statements and no later than 110 days after the end of the fiscal year. Notably, these reports must include an assessment by the Board of Management on the company's environmental and social responsibility. (Circular 96/2020/TT-BTC, Annex 4). According to Article 7 of Circular 96/2020, public companies are required to disclose their annual reports on all of the following channels: a) Their website; b) The media of the State Securities Commission of Vietnam; c) The website and other information disclosure media of Stock Exchange according to its regulations; d) Website of Viet Nam Securities Depository and Clearing Corporation (VSDCC); dd) Other means of mass media as prescribed by law (printed newspapers, online newspapers, etc.). This ensures that all stakeholders of business enterprises can access the report and assess whether and how the enterprises fulfill their CSR. Non-compliance with this requirement or the disclosure of inaccurate information may result in administrative penalties of up to VND 200 million (approximately USD 8,100). In addition to monetary fines, violators may also be required to correct any inaccurate disclosures (Decree 128/2021/ND-CP, Article 1, Clause 34). Clearly, the role of the State in establishing the CSR policy framework and promoting CSR is becoming increasingly important.

# 3 ISO 26000: 2010

The concept and specific contents of CSR have been codified into norms and standards for worldwide applications. One of the most well-known and core international standards for CSR is ISO 26000: 2010 (ISO 26000) which was drafted by the Working Group on Social Responsibility of International Organization for Standardization, the world's largest developer of voluntary international standards.

ISO 26000 provides voluntary guidance; there are no requirements, and thus it is not used as a certification standard. Its scope applies to organizations of all types, both in the public and private sectors of all types and sizes, in developed and developing countries. As environmental CSR is one of the core CSR subjects, ISO 26000 makes suggestions on how to identify environmental problems and mitigate the environmental impact of business enterprise activities.

ISO 26000 looks into the following issues: (1) prevention of pollution; (2) sustainable resource use; (3) climate change mitigation and adaptation; and (4) protection of the environment, biodiversity, and restoration of natural habitats. These issues are elaborated in the following analytical framework:

## *3.1 Prevention of Pollution (ISO 26000, 6.5.3)*

In order to improve the prevention of pollution caused by its activities, businesses need to (i) identify aspects and impacts of the organization's decisions and activities on the surrounding environment; (ii) identify sources of pollution and waste related to the organization's operations; taking measures to prevent pollution and waste, employing waste management systems; and ensuring proper management of pollution and waste that can be avoided; (iii) engage with local communities on actual and potential polluting emissions and wastes, related health risks, as well as practical and proposed mitigation measures; (iv) disclose the amounts and types of relevant and significant hazardous materials used and discharged; (v) avoid the use of banned chemicals such as ozone depleting substances, durable organic pollutants, dangerous chemicals, pesticides, cancer-causing chemicals, mutagenic, reproductive effects, disorder endocrine disturbances, or long-term and toxic bioaccumulation.

## *3.2 Sustainable Resource Use (ISO 25000, 6.5.4)*

Environment—identifying and improving environmental impacts of the operations, including resource use and waste disposal. In order to sustain resources in the future, current sizes, volumes of consumption, and production need to change so that they operate within the Earth's capabilities. The sustainable use of renewable resources means that they are used at a rate less than or equal to nature's addition to nonrenewable resources (such as fossil fuels, type, minerals), long-term sustainability requires a rate of utilization less than the proportion of replaceable resources. The organization can move toward sustainable resource use by using electricity, fuel, and raw materials more responsibly, as well as by using processed materials, land, and water more efficiently.

### *3.3 Climate Change Mitigation and Adaptation (ISO 26000, 6.5.5)*

Emission of the greenhouse gases (GHG) such as carbon dioxide, methane, and nitrogen oxides has a severe effect on the environment and human health. All organizations need to be responsible for the development of greenhouse and will be affected by climate system change in a certain way.

To mitigate the impacts of climate change related to their activities, businesses need to identify sources of direct or indirect GHG emissions and define the scope of their responsibilities; measuring, documenting, and reporting on a GHG development organization that is worthwhile; taking optimal protective measures to gradually reduce direct and indirect GHG emissions within the organization's control and to encourage similar activities within the organization; realizing energy saving; take countermeasures against valid or contingent activities and contribute to the work of enhancing the adaptive capacity of stakeholders within their sphere of influence.

### *3.4 Protection of the Environment, Biodiversity, and Restoration of Natural Habitats (ISO 26000, 6.5.6)*

Business enterprises can become more socially responsible by working to protect the environment and restore natural habitats and many other functions and services the ecosystem provides (such as food and water, climate regulation, soil formation, and regeneration opportunities). Aspects of this include valuing and protecting biodiversity; protecting, and restoring ecosystem services; sustainable use of land and natural resources; urban and rural development with good environment.

## 4 Formosa Ha Tinh Steel Corporation Case

In this part, the authors bring up the case of Formosa Ha Tinh Steel Corporation's incident in 2016 to provide an account on (i) how business enterprises, typically foreign invested company respond to environment preservation in Vietnam, (ii) what actions and policies Vietnamese government has taken to regulate environmental issues and impacts caused by corporate sector, and (iii) how international law and domestic law on CSR has been violated.

Due to the limited public information availability, the authors do not have full access to the details of each violation. Hence, this chapter will provide information and opinion of the authors publicly concluded by the government on mainstream newspapers.

### *4.1 History of Formosa Steel Corporation*

Formosa Ha Tinh Steel Corporation was founded by Formosa Plastics Group (FPG), a Taiwanese corporation set up in 1954. With more than 65 years of development, FPG is now one of the world's foremost plastics companies and the largest private enterprises in Taiwan (D&B Business Directory, 2020). The group also has its global presence in the United States, China, Vietnam, the Philippines, Indonesia, and other regions.

### *4.2 Historical Environment Issues of Formosa Steel Corporation*

Since its establishment, Formosa Steel Corporation has faced several legal suits mainly for causing environmental pollution not just in Taiwan but also in the host countries where their branches, subsidiaries, and other commercial presence are located around the world (Violence Tracker, 2020).

The corporation has shown a "systematic" lack of responsibility for polluting the environment in which it invests (Table 1).

Among the cases, a notable case was in relation to their subsidiaries, in 2009, Formosa Plastics Corp., Texas, and Formosa Plastics Corp., Louisiana in the United States which were accused of violating the regulatory requirements on leak prevention/controls of air pollution. Consequently, the penalty applied to these subsidiaries was $12,800,000 (Violence Tracker 2020).

Recently, Texas residents filed a lawsuit against Formosa Plastics Corp. for discharging billions of plastic pellets into Lavaca Bay and Cox Creek, leading to the pollution of Texas Waterways. The case was settled when the corporation agreed to pay $50 million, undertaken to resolve the existing pollution, and committed not to discharge plastics in the future ("Formosa Agrees to Pay Settlement," 2019).

It is notable that in 2009, the corporation was granted the infamous award "Black Planet" for their repeated violations to the environment (Canfield, 2020). This award is annually given to the world's most egregious polluters who "have committed

**Table 1** Primary offense types

| Primary offense types | Penalty total | Number of records |
|---|---|---|
| Environmental violation | $19,916,707 | 35 |
| Workplace safety or health violation | $768,599 | 22 |
| Railroad safety violation | $105,800 | 12 |

Source: Violation Tracker (2020)

themselves to the destruction and downfall of our Blue Planet in an outstanding way" (Wild at Heart Legal Defense Association, 2010).

# 5 Formosa in Vietnam

Hung Nghiep Formosa Ha Tinh Steel Company Limited (FHS) and the investment project were duly licensed in 2008 in accordance with the Investment Certificate No. 282023000001 to carry out the following objectives: Metallurgical, production, trading and export of iron and steel; port business; produced from kilns; cement, pressed products. With the total investment capital of USD7,879,060,000, in which the charter capital contributed by the investor, Formosa Plastics Group, is USD 2,700,000,000. The project is located in Ha Tinh province, Vietnam.

In 2012, the total investment capital of the project was increased to USD22 billion. Their refinery plant was capable of producing and supplying four million tons of products annually. The Vietnamese government expected the project would create 30,000 jobs for local employees upon its official commencement of operation ("Formosa lifts investment in cast iron refinery," 2012).

## *5.1 The Incident and Irresponsible Actions of Formosa*

### 5.1.1 The Incident

In early April 2016, marine environmental pollution "revealed" from the fish death phenomenon in the Vung Ang seaport area in Ky Anh town, Ha Tinh province where the investment project of FHS is located ("First death in Formosa toxic leak," 2016).

Fishery mortality phenomenon spread on a large scale, starting in the coastal area of Ha Tinh, then spreading along the coastal provinces of Quang Binh, Quang Tri, and Thua Thien-Hue.

This incident has caused serious economic, social and environmental damage, of which the fishery sector is most heavily affected, followed by business, service, tourism and daily life of fishermen.

The cause of the environmental incident causing mass death of seafood in the coastal provinces of four central provinces was determined by FHS during the test operation of the factory complex, there were violations and to happen. Causing incidents, resulting in untreated wastewater containing phenol toxins and cyanide discharged into the environment.

In late June 2016, the government official concluded that FHS was responsible for this marine ecological disaster, as the company discharged untreated wastewater into the sea.

During the inspection at FHS, the Government also detected and found that FHS breached 53 procedural regulations, mainly in respect to the design, operation, construction, and implementation of the plant. More seriously, FHS illegally (without prior approval of competent authorities) changed the coke dry quenching waste treatment technology (environmentally friendly technology) to the coke wet quenching waste treatment one which generates more waste, especially emissions into the environment (Phu, 2016).

In July 2016, FHS was detected to illegally bury 100 tons of stinky black waste at a farm in Ky Trinh ward ("Formosa found burying waste again", 2016).

In December 2017, FHS was again detected violating environmental regulation on classifying and determining the quantity and weight of hazardous waste for management (Decree 155/2016/ND-CP, Art 21.5.c) and transferring, giving, or selling hazardous waste to an organizational or individual entity that is not licensed to treat hazardous waste (Decree 155/2016/ND-CP, Art 21.7.h) in accordance with Decision 3745/QD-UBND issued by the Ha Tinh People's Committee. Consequently, FHS was subject to an administrative penalty of VND560 million ("Formosa fined for improper waste treatment," 2017).

## 5.2 Actions Taken by FHS

In response to the incident and question about "why FHS let this accident happen" raised by VTC14 television channel, Mr. Chou Chun Fan, Deputy Head of Formosa's Public Relations Department, Head of the Formosa Office in Hanoi caused a "storm" of public opinion when providing a shocking statement that "It is impossible to build a steel plant here without leaving any impact on fish and shrimp… Between these two things, we must choose one?" ("Formosa under fire over mass fish deaths", 2016).

After being criticized by the public and granted with the government decision on its violation, FHS admitted to causing the mass fish deaths, publicly apologized to the Vietnamese government and local people, and pledged to pay $500 million in compensation for the damages ("Formosa owns up to fish kill disaster", 2016).

On 30 August 2016, Formosa completed transferring compensation money to the Vietnamese Government with the total amount of USD500 million as committed. ("Results of solving marine environmental problems," 2017).

## 5.3 *Impacts*

- About 70 tons of dead fish were found in the four provinces, affecting more than 270,000 people who earn their living from the sea (VNS, 2017). According to the preliminary calculation, the environmental pollution incident caused by FHS directly affected more than 100 thousand people due to lack of stable jobs, low income and 176,285 dependents; Loss of marine and coastal catches are estimated at 1600 tons per month (Han, 2016).
- In addition, environmental disasters and natural disasters are also one of the main factors influencing Vietnam's growth. Due to the impacts caused by the pollution, the national GDP suffered a loss of 0.3–04% growth rate (Manh, 2016). According to the target, Quang Binh province's GDP growth rate in 2016 would reach 8%, but due to the Formosa incident, the growth rate was only about 4.2% (Ho, 2016).
- According to governmental statistics, Ha Tinh province has more than 24,400 people affected by the crisis, of which more than 15,000 people are directly engaged in fishing, aquaculture, the rest are fisheries and logistics services. 1000 people, nearly 700 people in hotel and restaurant service, 428 people in salt production.
- Affecting Vietnam's attraction of FDI, as the Vietnamese government would be stricter in reviewing all foreign invested projects. Meaning foreign investors may face more procedural challenges when registering/establishing their investment projects, especially those which may have environmental impacts before issuing licences.
- More significantly, the incident created not only eco-social problems, but also issues of security and national defence (Tuoi Tre News, 2016).

## 5.4 *Did Formosa Follow the International CSR Standards and Vietnamese Law*

Per the above analysis on CSR concept and the Formosa case study, it is undeniable that business enterprises are responsible for society. Environmental social responsibility is a part of CSR, demonstrating commitment to managing and improving the environmental impacts created by business activities.

Considering FHS activities and violations in Vietnam, the authors found that both parent company and FHS failed to carry out CSR required by the international CSR standards and Vietnamese law.

Failure in conducting international guidance on CSR:

(a) Disrespecting ISO 26000

According to ISO 26000 (6.5.2.1), the responsibility for protecting the environment is a must. In addition to the compliance with regulations, business enterprises are required to admit their responsibility for the environmental impacts caused by their operations. However, not until the case of illegal burying 100 tons of stinky black waste at a farm in Ky Trinh ward was investigated and applied responsibility to FHS, the company had rejected an accusation. This violates the first principle of environmental responsibility.

Environmental risk management principle requires business enterprises to adopt programs using risk-based and sustainability-based perspectives to assess, avoid, and mitigate environmental risks and impacts of their activities. Additionally, they also need to raise awareness and provide information on environmental incidents to authorities and local communities (ISO 26000, 6.5.2.1). Nevertheless, from the answer of their spokesperson that you cannot have both steel plant and seafood, it is very much likely that awareness about CSR is not existing in the company.

Another principle stating that enterprises must bear all costs to settle the pollution affected by society (ISO 26000, 6.5.2.1). One issue to note here is that the figure of USD500 million is on what basis and criteria to set the amount of USD500 million and whether this number is enough to compensate the people and the marine environment in Vietnam. Or not, if it is not sufficient, how will it be settled? In addition, the time of compensation and remediation also needs to be specified in detail and clarity. Per the above section, there were 270,000 people directly affected by this violation, meaning each individual (not including business enterprises) is only entitled to USD1850 which is sufficient from the authors' view.

(b) Bad CSR

Applying the Carroll's CSR pyramid, FHS is subject to four levels of CSR including economic, legal, ethical, and philanthropic responsibilities.

At an economic level, if FHS makes a loss, it will be considered a "failure" to exercise this responsibility. In 2019, FHS achieved revenue of VND 71,664 billion, an increase of 11.66% compared to 2018 (64,175 billion). However, the company's profit after tax incurred a loss of VND11,538 billion in 2019, and a loss of VND2,727 billion in 2018. Thus, FHS has already accumulated losses amounting VND25,388, equivalent to nearly USD1.1 billion USD (Cao, 2020).

Assessing the second level, both international law and domestic laws have strictly regulated CSR in various legal documents. Protecting the environment and reducing resource consumption are two important issues in CSR. These laws aim to force businesses to increase their environmental accountability by building a closed process, ranging from resource handling and development, product design, assembly, marketing, and consumption and recycling. Nevertheless, by continuously violating several commitments and policies (e.g., 53 violations during the inspections and other violations after the incident) stipulated by the Vietnamese government, it is very much likely to confirm that FHS has entirely failed to perform legal responsibilities. Importantly, after the incident, FHS has not shown signs of starting to comply with the laws, the company keeps breaching the rules while illegally burying hazardous waste.

Regarding two highest levels of CSR, the authors have not found any evidence or information proving that FHS has exercised these responsibilities.

Failure in fulfilling its investment commitments and violations with Vietnamese law:

Per the above information in relation to the violations of FHS, the company was accused of breaching 53 regulatory requirements in terms of environmental protection.

Furthermore, the fact that Formosa has caused the death of fish has signs of violation under the 1999 Penal Code (amended in 2009) in Article 182 Crime of causing environmental pollution and Article 188 Crime of destroying aquatic resources. However, to determine whether there is a criminal violation in this case or not, it is necessary for the competent authorities to conduct an investigation to clarify the violations, the extent of violation in this case to be able to be handled properly, in compliance with the provisions of the law.

## 6 Conclusion

After a brief review of the CSR concept and ISO 26000 guidance on environmental CSR, the chapter analyzes the Hung Nghiep Formosa Ha Tinh case to withdraw lessons learned, from which the recommendations are made.

### *6.1 Raising Awareness on CSR*

It is important to note that the first step of applying CSR to protect the environment comes from the understanding of each individual about the importance and impact of CSR on business sustainability and other stakeholders. The understanding can only be perceived correctly, if investors, corporate managers and staff fully understand (i) what responsibilities they have; (ii) why they are required to exercise CSR; and (iii) how they can exercise CSR.

Looking back to the case of FHS, it is clear that even the mindset of the corporation acted irresponsibly and showed a lack of understanding about CSR on the stakeholders when saying we must choose either "fish or factory." This is totally incompliant with both international law and domestic law regulating CSR. Hence, the first recommendation to make for both business enterprises and lawmakers is to raise certain awareness about CSR. Specifically, business enterprises should provide more information and organize more training sessions for employees about the role and importance of CSR to business sustainability and environment. Even when it is proved that the country has issued several legal documents regulating CSR on environment as analyzed above, the authors found that strong communications on CSR is missing. This may be a rationale for FHS and other similar cases which keep accumulating. Knowledge and understanding should firstly come from education. Therefore, the Ministry of Education and Training should consider issuing necessary circulars and decisions regulating CSR/responsible business and management/

sustainability/business ethics as a compulsory module at higher education level. Alternatively, the Ministry of Education and Training can consider requiring the undergraduate programmers in economics and management disciplines (e.g., business administration, public management, and human resource administration) to include this module. This will help raise awareness of students who will then be future workers, investors to manage and practice CSR, and promote the sustainable development of businesses in Vietnam.

More importantly, it is recommended that the educational modules in relation to CSR and training sessions in environmental aspects should include the contents answering to the question "why business enterprises need to implement CSR" and highlighting the benefits of environmental protection to business enterprises. Firstly, they should understand that if business enterprises are willing to seek more and better output, they must have their inputs stabilized. If the natural resources used for production are damaged, they should expect costs of production to be higher. Additionally, several rationales mentioned in the works of "noneconomic approach" such as Bowmen's and Johnson's should be good answers to the questions why we need to implement CSR and its benefits.

## 6.2 Setting Up a Stronger Legal System to Manage CSR

Rapid growth and the natural environment are always two different sides of the coins that developing economies need to consider. If the country paid too much focus on rapid economic growth without considering other factors, several consequences would be emerging. The case of FHS is a clear example for this sacrifice as a price for rapid growth in the short term.

From the study case, the management of waste and hazardous waste emerges as one of the problematic issues that need to be resolved. In general, Vietnam has issued a relatively complete system of legal documents to enforce the implementation of CSR in the environmental area. In which, responsibilities and rights of state agencies, individuals, and organizations conducting activities with respect to environmental issues and sanctions for handling violations are specified.

However, the authors found that there are still some limitations and gray areas in legislation that should be improved in this regard. As one of the main drivers of CSR implementation, the authors of this paper opines that a strong legal reform regulating CSR in environmental protection area needs to be implemented as follows:

## 6.3 Enhancing the Assessment Procedures of Investment Projects Having Impacts on Environment

As per the above assessment of Law on Investment, to be granted with the IRC, the investors must submit the preliminary assessment of environmental impacts and environmental protection measures. However, pursuant to the former Deputy Minister of Planning and Investment, at the time when the ministry assessed the application dossier submitted by the investor, it has provided a comment that "The environmental impact assessment of the project is still sketchy/simple, does not contains factors such as the source of the impact, the affected object and scale, impact price, negative impact mitigation measures, prevention methods of environmental incidents. Investors are requested to prepare environmental impact assessment reports and submit them to competent state agencies for consideration and approval according to current regulations" ("Lesson from Formosa," 2016). Nevertheless, it is likely that FHS was still granted a license for investment projects without meeting the requirements set by the Ministry of Investment and Planning. This indicates two issues which are (i) the regulated form of environmental impact assessment report is not detailed enough and (ii) opinion of the reviewing state agency is not carefully considered in the licensing process.

Hence, an amendment to the current standard form of environmental impact assessment report and a revision of current regulations to include compulsory explanatory responsibility of investors based on the comments of competent State agencies are recommended.

## 6.4 Strengthening Regulations on Waste Management and Exchange

Waste management practices in many countries around the world have also shown the economic and environmental value of waste exchange between enterprises. Accordingly, this enterprise's waste can become a source of raw materials for one or more other businesses. Therefore, the exchange of wastes between these groups of enterprises not only saves production costs but also has significant value in preventing environmental pollution or overexploitation of natural resources. In that sense, it is necessary to add regulations on waste exchange to the current Law on Environmental Protection to create a legal basis for waste exchange between enterprises, thereby maximizing economic and environmental value while ensuring benefits for businesses.

## 6.5 Providing Detailed Regulations on Hazardous Waste Transportation and Responsibilities of Waste Treaters

In fact, there are still some legal provisions related to hazardous waste transportation that are general and not feasible, such as:

Under the provisions of Decree 38/2015/ND-CP (Art 12.1) on management of waste and discarded materials, the hazardous waste treaters are responsible for signing a contract with the source owners to collect, transport, and treat hazardous waste. They must dispose the hazardous waste within the operational area recorded in their hazardous waste treatment license of and ensure that the quantity and types of hazardous wastes are received, transported, and treated using permitted means, systems, and equipment, in accordance with the contract, hazardous waste documents, and hazardous waste disposal license. While Circular 36/2015/TT-BTNMT (Art 9.2) stating that the hazardous waste treater shall take measures to manage and take full responsibility for the operation of the non-primary means of transport, which is not owned in the process of transporting hazardous waste; report to the Vietnam Environment Administration on the change of contents, extension or termination of the contract for handing over the non-governmental vehicle within 15 working days from the date of change, extension or termination.

Thus, the regulations on waste treaters on waste transportation just stop at "taking full responsibility for the operation of the unofficial means of transport in the process of transporting hazardous waste," but how much responsibility the waste treaters need to take, is not available.

To regulate the transportation of vehicles of hazardous wastes, the law requires such vehicles must be equipped with Global Positioning Systems (GPS) which is always connected to the online network to search and record the travel of hazardous wastes (Circular 36/2015/TT-BTNMT, Art 8.2). Failing to comply with this requirement may impose an administrative penalty to business enterprises up to VND100 million (Decree 155/2016/ND-CP, Art 22.2 and Art 22.3). Nevertheless, due to the current economic and technology development in Vietnam, the authors trust that this regulation is not entirely practical. Particularly, the Internet connection is not always available or unstable in rural areas, leading to the fact that recording and locating the travel of hazardous waste would not be possible. Hence, it is important to amend the regulations on how to use the GPS for vehicles appropriately, so that they can be effectively applied in practice. This should be implemented together with telecommunication development and issuance of specific regulations and guidance on how to use GPS.

## 6.6 Enhancing Responsibilities and Power of Environmental Agency

Although the Ministry of Natural Resources and Environment issued Decision 1659/QD-TNMT dated July 22, 2016 establishing the Supervisory Team to monitor FHS's activities ("Supervisory Team of Formosa," 2016), the authors are of the opinion that this should have been done from the beginning of the project, not to wait until the incident happened.

Thus, it is recommended that the government should have a specific state management agency on environmental protection to monitor the projects with high environmental risks.

## 6.7 Enforcing Penal Codes on Environmental Polluters

In the case of FHS, there have not been any legal proceeding procedures on criminal signs caused by FHS. In accordance with the Penal Codes (2015), there are several offences on (i) causing environmental pollution (Art 325), (ii) hazardous waste management (Art 326); (iii) prevention, response, and relief of environmental emergencies (Art 327); and (iv) destruction of aquatic resources (Art 342).

Nevertheless, business enterprises are not subject to the offence of hazardous waste management (Art 2 and Art 76.2). It is recommended to supplement in the criminal liability of the juridical/legal person for the crime for hazardous waste management in order to avoid leaving *criminals*.

## 6.8 Setting Up Whistleblowing System at Corporate Level

Given that the law is in place and business enterprises are willing to implement both regulatory and voluntary CSR; however, it is necessary for them to have an internal group/team to watch out and prevent irregular and noncompliant CSR activities.

Hence, it is suggested having a system and allocating personnel in charge of listening to the opinions of each stakeholder and assessing information quality, reporting to the management board and responding to the stakeholders.

In conclusion, it is important to reemphasize that companies hold responsible for environmental problems; CSR implementation is a must for every business enterprise, otherwise, cases similar to FHS will be occurring in the future. Ensuring the effective implementation of CSR needs efforts from the state to regulate and supervise, from businesses to comply with both regulations and voluntary guidance, such as ISO 26000 on CSR, and from other stakeholders to observe, assess, and reflect the implementation of CSR. Although the current policies on CSR in Vietnam are strengthened by adopting international law and have changed positively to regulate environmental issues, it needs to be improved to exercise its role in guiding, binding, and encouraging business enterprises to exercise CSR.

# References

## Legal Documents

Circular 96/2020/TT-BTC, Appendix 04
Decree 128/2021/ND-CP, Art 1.34
Circular 36/2015/TT-BTNMT, Art 9.2
Companies Act 2006, s 172
Companies Act 2013, s 135
Decree 155/2016/ND-CP, Art 21.5.c
Decree 155/2016/ND-CP, Art 21.7.h
Decree 155/2016/ND-CP, Art 22.2 and Art 22.3
Decree 38/2015/ND-CP, Art 12.1
Directive 2014/95/EU, Art 3
Directive 2014/95/EU, Art 5
Directive 2014/95/EU, Art 6
ISO 26000, 6.5.3
ISO 26000, 6.5.4
ISO 26000, 6.5.5
ISO 26000, 6.5.6
Penal Codes 2015, Art 235

## Articles and Books

Bergin T (2014) Starbucks suffers first UK sales drop after tax criticism. Reuters. Retrieved November 16, 2020 from https://uk.reuters.com/article/uk-starbuckstaxbritain-idUKBREA3N0X020140424
Carroll AB (1991) The pyramid of corporate social responsibility: toward the moral management of organizational stakeholders. Bus Horiz 34(4):39–48
Carroll AB (1999) Corporate social responsibility: evolution of a definitional construct. Bus Soc 38(3):268–295
European Commission (2001) Promoting a European Framework for Corporate Social Responsibility – Green Paper. Publications Office. 366, at 6
Friedman M (1970) The social responsibility of business is to increase its profits. The New York Times Magazine
Johnson HL (1971) Business in contemporary society: framework and issues. Wadsworth Publishing Company, p 50–51
Jowit J (2010) World's top firms cause $2.2tn of environmental damage, report estimates. The Guardian. Retrieved November 14, 2020 from https://www.theguardian.com/environment/2010/feb/18/worlds-top-firms-environmental-damage
Matten D, Moon J (2004) "Implicit" and "explicit" CSR: a conceptual framework for understanding CSR in Europe. International Centre for Corporate Social Responsibility
McGuire JW (1963) Business and society. McGraw-Hill, New York, p 144
Smith RE (2011) Defining corporate social responsibility: a systems approach for socially responsible capitalism. University of Pennsylvania, Philadelphia, PA
Steiner GA (1971) Business and society. Random House, New York, p 164
Wood DJ (1991) Corporate social performance revisited. Acad Manag Rev 16(4):691–718

## Others


Asia Nikkei (2016) First death in-suspected Formosa Plastics toxic leak. Retrieved November 20, 2020 from http://asia.nikkei.com/Business/Companies/First-death-in-suspected-Formosa-Plastics-toxic-leak

Bao Cao Mau (2017) Results of solving marine environmental problems in some central provinces. Retrieved November 28, 2020 from http://www.baocamau.com.vn/sinh-hoat-tu-tuong/ket-qua-giai-quyet-su-co-moi-truong-bien-mot-so-tinh-mien-trung-46445.html> accessed 28 November 2020

Canfield S (2020) Massive plastics plant granted air quality permits for construction in Louisiana. Courthouse News Service. Retrieved November 20, 2020 from https://www.courthousenews.com/massive-plastics-plant-granted-air-quality-permits-for-construction-in-louisiana/

Cao C (2020) Formosa Ha Tinh has accumulated losses of more than 1 billion USD after 2 years of operation. Cafef. Retrieved November 30, 2020 from https://cafef.vn/formosa-ha-tinh-da-lo-luy-ke-hon-1-ty-usd-sau-2-nam-di-vao-hoat-dong-20200831183839325.chn#:~:text=Theo%20s%E1%BB%91%20li%E1%BB%87u%20c%E1%BB%A7a%20ch%C3%BAng,2018%201%E1%BB%97%202.727%20t%E1%BB%B7%20%C4%91%E1%BB%93ng

CTTDT. Establishing the Supervisory Team of Formosa Ha Tinh Company in implementing commitments to overcome environmental incidents in 04 central provinces. Trang thong tin Bo truong Bo Tai nguyen va Moi truong. Retrieved November 30, 2020 from http://botruong.monre.gov.vn/tin-tuc/chi-dao-dieu-hanh/thanh-lap-to-giam-sat-cong-ty-formosa-ha-tinh-trong-viec-thuc-hien-cam-ket-khac-phuc-su-co-moi-truon-2855

D&B Business Directory (2020) Formosa plastics corporation – Company profile. Retrieved November 18, 2020 from https://www.dnb.com/business-directory/company-profiles.formosa_plastics_corporation.29cc1714e46499c2f5a90b4fecc2fe00.html

Han V (2016) More than 276,000 people were affected after the Formosa discharge. Cong An Nhan Dan Online. Retrieved November 28, 2020 from http://cand.com.vn/doi-song/Hon-276-000-nguoi-bi-anh-huong-do-Formosa-401057/

Ho L (2016) The Formosa case reduced Quang Binh's GDP by half. Viet Times. Retrieved November 28, 2020 from https://viettimes.vn/vu-formosa-lam-giam-mot-nua-gdp-cua-quang-binh-post41451.html

Manh N (2016) The Formosa environmental incident caused a loss of 0.3% of Vietnam's GDP. Thiennhien.net. Retrieved November 28, 2020 from https://www.thiennhien.net/2016/12/28/su-co-moi-truong-formosa-gay-thiet-hai-0-3-gdp-cua-viet-nam/

Phu T (2016) Formosa có tới 53 vi phạm (Formosa has 53 violations). Phap Luat Thanh Pho Ho Chi Minh. Retrieved November 25, 2020 from https://plo.vn/thoi-su/formosa-co-toi-53-vi-pham-640146.html

Plastic Pollution Coalition (2019) Formosa plastics agrees to pay $50 million settlement for polluting Texas waterways. Retrieved November 20, 2020 from https://www.plasticpollutioncoalition.org/blog/2019/10/15/formosa-plastics-agrees-to-pay-50-million-settlement-for-polluting-texas-waterways

Straitstimes (2016) Taiwan firm Formosa under fire over allegations of pollution in Vietnam mass fish deaths. Retrieved November 25, 2020 from https://www.straitstimes.com/asia/se-asia/taiwan-firm-formosa-under-fire-over-allegations-of-pollution-in-vietnam-mass-fish

Thanh Nien News (2016) Formosa unit owns up to fish kill disaster, commits to $500 million compensation. Retrieved November 28, 2020 from http://www.thanhniennews.com/society/formosa-unit-owns-up-to-fish-kill-disaster-commits-to-500-million-compensation-63688.html#:~:text=June%2030%2C%202016-,Formosa%20Ha%20Tinh%20Steel%20Corp.,Vietnamese%20government%20said%20Thursday%20afternoon.&text=The%20company%20committed%20to%20apologize,the%20disaster%20and%20pay%20VND11

Thoi Bao Tai Chinh Viet Nam (2016) Lesson learnt from Formosa case: there is no trade-off in the environment to attract foreign investment. Retrieved from http://thoibaotaichinhvietnam.vn/

pages/kinh-doanh/2016-07-01/bai-hoc-tu-formosa-khong-danh-doi-moi-truong-de-thu-hut-dau-tu-nuoc-ngoai-33135.aspx
Tuoi Tre News (2016) Formosa environmental scandal in Vietnam poses potential security problems. Retrieved November 28, 2020 from https://tuoitrenews.vn/news/society/20160713/formosa-environmental-scandal-in-vietnam-poses-potential-security-problems/16225.html
Tuoitrenews(2016)Formosatakesresponsibility for Vietnam's fish deaths, pledges $500mn in compensation. Tuoi Tre News. Retrived September 15, 2020 from: https://tuoitrenews.vn/society/35634/formosa-takes-responsibility-for-vietnams-fish-deaths-pledges-500mn-in-compensation
Vietnamnews (2020) Formosa Steel lifts investment to $22b in cast iron refinery. Retrieved November 20, 2020 from https://vietnamnews.vn/economy/223894/formosa-steel-lifts-investment-to-22b-in-cast-iron-refinery.html
Violation Tracker (2020a) Violation tracker individual record – Formosa plastic corporation. Retrieved November 20, 2020 from https://violationtracker.goodjobsfirst.org/violation-tracker/-formosa-plastics-corp
Violation Tracker (2020b) Violation tracker parent company summary – Formosa plastic corporation. Retrieved November 20, 2020 from https://violationtracker.goodjobsfirst.org/parent/formosa-plastics
VNS (2016) Formosa found burying waste again. Vietnamnews. Retrieved November 25, 2020 from https://vietnamnews.vn/society/299619/formosa-found-burying-waste-again.html
VNS (2017) PM warns Formosa against future environmental violations. Vietnam News. Retrieved November 28, 2020 from https://vietnamnews.vn/environment/380718/pm-warns-formosa-against-future-environmental-violations.html
VOV Giao Thong (2017) Burial of hazardous waste originating from Formosa Ha Tinh: a fine of more than VND1 billion. Retrieved November 25, 2020 from https://vovgiaothong.vn/Chon-lap-chat-thai-nguy-hai-co-nguon-goc-tu-Formosa-Ha-Tinh-Phat-hon-1-ty-dong
Wild at Heart Legal Defense Association (2010) Save the Taiwanese white dolphins. Taiwansousa Blogspot. Retrieved November 20, 2020 from http://taiwansousa.blogspot.com/2010/05/black-planet-award-for-fpg.html
World Bank (2023) GDP growth (annual %) – Vietnam. Retrieved March 10, 2025 from https://data.worldbank.org/indicator/NY.GDP.MKTP.KD.ZG?locations=VN
WTO (2007) Viet Nam and the WTO. Retrieved September 15, 2020 from https://www.wto.org/english/thewto_e/countries_e/vietnam_e.htm

**Minh-Duc Le** is a LLM holder with distinction graduating from the program of Master of Laws in International Trade and Economic Law, a collaborative program between Foreign Trade University and the University of the West of England. He currently holds two positions at British University Vietnam (BUV), including compliance manager and a member of the Senate. Minh has 15 years of working experience in the law sector encompassing 6 years working for PwC Legal Vietnam and 9 years working for BUV. His specialized expertise is law on education, law on investment, and law on enterprises.

**Assoc. Prof. Hien-Thi Tran** (Ph.D in Management, Southampton University Business School, UK) has research interest in corporate social responsibility, business ethics, corporate culture, corporate governance, and human resources. She works with the VNU University of Economics and Business, Vietnam National University Hanoi. Dr. Tran has published in CABS journals such as *Journal of Business Research*, *Multinational Business Review*, *Corporate Governance: The International Journal of Business in Society*, and *International Journal of Business Governance and Ethics*. She is the lead editor of an editorial team for a Palgrave MacMillan book indexed in Scopus.

# Social Enterprise Development for Sustainability in Vietnam

**Trang-Thu Tran**

**Abstract** Social enterprise has been increasingly acknowledged as a new organizational model to address social and environmental issues and thus promote sustainability around the world. Broadly defined, social enterprises are organizations that pursue a social mission through commercial activities. Being the most recent forms of organization, they often lack legitimacy and resources to succeed at significant scale. For social enterprises to scale up, an enabling environment that includes appropriate funding, new legal forms, and the assimilation of the values associated to this organizational model is necessary. In Vietnam, several local and international organizations have sought to create such an environment for social enterprise development. As a result, social enterprise has been recently recognized in the Enterprise Law 2014. This chapter examines the activities of the first organizations that aim at promoting the diffusion and development of social enterprises in the country.

## 1 Introduction

Large-scale economic reforms in 1986 have transformed Vietnam from one of the poorest countries in the world to a lower middle income country with real GDP per capita from less than US $700 in 1986 to almost US $4,500 in 2023 (World Bank, 2024). Indeed, there has been a considerable reduction in poverty headcount ratio at societal poverty line from 49.8 percent of the population in 1992 to 19.9 percent in 2022 (World Bank, 2025). Despite remarkable economic achievements, Vietnam continues to struggle with many social challenges such as aging population, large rich poor gap, unequal access to public health and education, and severe environmental pollution. Due to resource constraints and weak capacity in providing social welfare, the government is unable to address all these issues. Meanwhile, being

T.-T. Tran (✉)
Foreign Trade University, 91 Chua Lang Street, Ha Noi, Vietnam
e-mail: thutrang@ftu.edu.vn



H.-T. Vu, L.-P. Dana (eds.), *Vietnam on the Road to Sustainable Economic, Social and Business Development*, Frontiers in South and Southeast Asian Development Research, https://doi.org/10.1007/978-981-96-4371-4_12

driven by monetary profit, private for-profit enterprises have frequently ignored the needs of the most disadvantaged groups in society. Nongovernmental organizations (NGOs) with the core mission of benefiting the community are expected to fill the gaps that neither for-profits nor the government have satisfied. However, these organizations have relied on international grants for years. As the country achieved lower middle income status, the source of funding to NGOs started to reduce, leading many NGOs to look for alternative ways to pursue their mission and be financially sustainable at the same time. Since 2009, several NGOs have introduced social enterprise as a new organizational model to solve social issues and promote sustainability in Vietnam (Nguyen et al., 2012). Research has estimated that there are 19,125 organizations of this type in Vietnam, with the majority being focused on work integration and being key employers of women and young people (British Council, 2019). Despite their existence and contributions to society, social enterprises have not been widely acknowledged by the public. In addition, there was a lack of State recognition and specific legal form for social enterprises before 2014, which made the creation and development of social enterprises challenging in Vietnam. For that reason, organizations such as British Council (BC) and Center for Social Initiatives Promotion (CSIP) have actively worked to shape an enabling institutional environment for social enterprise development in the country. This chapter will explore the work of these two organizations.

## 2 Literature Review

The social entrepreneurship field has grown rapidly in recent years around the world with numerous examples of dedicated and visionary social entrepreneurs working to satisfy unmet social needs (Bornstein, 2007; Pless, 2012). By contrast, social entrepreneurship research is still in an embryonic state: a unified definition is missing (Short et al., 2009) and scholars struggle to clarify the concept and delineate boundaries of the field (Brouard and Larivet, 2010; Dacin et al., 2010, 2011). For some authors, social entrepreneurship refers to "the activities" of individuals (Roberts and Woods, 2005) or collectives (Austin et al., 2006; Peredo and McLean, 2006) to create social value (Austin et al., 2006; Peredo and McLean, 2006) or social change (Roberts and Woods, 2005). Others regarded social entrepreneurship as "the process" starting from the identification of a social problem to the implementation of a solution to make social change (Mair and Martí, 2006; Martin and Osberg, 2007; Zahra et al., 2009; Bacq and Janssen, 2011). For example, Cochran (2007) defined social entrepreneurship as "a process of applying the principles of business and entrepreneurship to social problems" (p. 451). In addition, researchers suggest that social entrepreneurship is generally implemented through the creation of social enterprises that are "organizations seeking business solutions to social problems" (Thompson and Doherty, 2006, p. 362), regardless of their legal structures and degrees of financial self-sufficiency (Brouard and Larivet, 2010; Bacq and Janssen, 2011).

Previous research explores social enterprise as a new form of organization, focusing on its characteristics and inherent issues. Generally, social enterprises are defined as hybrid organizations that combine elements of both charity and business (Battilana and Lee, 2014; Ebrahim et al., 2014; Hervieux et al., 2010). Social enterprises are different from traditional NGOs and for-profit enterprises in their dual logics (the logic of nonprofit and that of for-profit) and identities (Battilana and Dorado, 2010; Moss et al., 2011; Tracey et al., 2011). Given the hybridity and liability of newness, social enterprises are not regarded as legitimate organizational forms, lacking resources for their operations (Battilana and Lee, 2014; Hervieux et al., 2010; Sud et al., 2009). For social enterprises to persist, an enabling institutional environment that includes a new legal form, financial support, and the assimilation of the values associated to this organizational form is needed (Dees, 2007; Florek, 2013; Sud et al., 2009).

Recent research on social enterprise has started to examine how social entrepreneurs and other actors in the field such as academics, foundations, and fellowship organizations create and diffuse the social enterprise model. However, these studies have mainly focused on discursive strategies to seek legitimacy and resources for social enterprises (Di Domenico et al., 2010; Hervieux et al., 2010; Nicholls, 2010; Ruebottom, 2013). Although prior research contributes to our understanding of how the social enterprise model is created and legitimized, it does not provide a detailed account of the strategies and activities to promote social enterprise development. Remarkably, little research has been done in the context of developing countries (Martí and Mair, 2009) where social enterprises can play a more important role in meeting the needs of socially disadvantaged groups. Therefore, more research is needed to understand the emergence and development of social enterprises in developing countries including Vietnam.

## 3 Theoretical Framework

To study the development of social enterprises in Vietnam, we use institutional theory, in particular the institutional work perspective (Lawrence and Suddaby, 2006; Lawrence et al., 2009). Institutional approaches to organization theory have traditionally focused attention on the relationships between organizations and the fields in which they operate, exploring the processes through which institutions govern action (Dimaggio and Powell, 1983; Meyer and Rowan, 1977; Zucker, 1977). However, the study of institutional work shifts the focus on understanding how action affects institutions (Lawrence et al., 2009). Institutional work is defined as "the purposive action of individuals and organizations aimed at creating, maintaining and disrupting institutions" (Lawrence and Suddaby, 2006, p. 215). Generally, the institutional work perspective highlights that individual and collective actors deliberately undertake actions to influence institutions in three important ways: creating, maintaining, and disrupting institutions. As social enterprise has emerged as an institutional phenomenon in Vietnam in recent years, the concept of institutional work is promising to unveil deliberate activities performed by the first organizations to promote the development of social enterprises. In this chapter, we focus on the process of institutionalizing social enterprise in Vietnam, in particular institutional work of social enterprise support organizations in the process.

# 4 Methods

We conducted a qualitative case study of the institutionalization process of social enterprise in Vietnam (2009–2014) in which local and international organizations were involved in institutional work to diffuse the social enterprise model. Although social entrepreneurship has occurred (at least in its cooperative form) in Vietnam, studies suggest that its modern form (i.e., social enterprise) did not develop before the establishment of the Center for Social Initiatives Promotion (CSIP) and the intervention of British Council Vietnam (British Council, 2011; Nguyen et al., 2012). CSIP is a Vietnamese nongovernmental and nonprofit organization with a sole mission of supporting the development of social enterprises in Vietnam. Since its establishment in 2008, CSIP has mainly worked at grassroots level, providing direct support to social enterprises at two stages of development: incubation and acceleration. To date, the center has nurtured and supported more than 80 social enterprises across Vietnam. Together with CSIP in the very first years of the Vietnam social enterprise movement, British Council (BC) was another key actor. BC is the United Kingdom's international cultural relations organization, which has offices in 110 countries and territories worldwide. BC has been in Vietnam since 1993 under the status of a foreign independent cultural and education and nonprofit organization. As an important part of its work in society, BC strongly promotes the development of social enterprises in the United Kingdom and around the world. BC launched its Global Social Enterprise Program in 2009 and currently operates this program in 24 countries. Together, BC and CSIP are pioneers in implementing the first social enterprise support program in Vietnam. The concept of social enterprise was first introduced by these organizations in 2009. Since then, more new social enterprises have been established (Nguyen et al., 2012). According to the first social enterprise mapping project commissioned by British Council (2011), 167 organizations were identified as social enterprises (although these organizations do not call themselves as such). However, recent research has estimated that there are 19,125 organizations of this type in Vietnam (British Council, 2019).

## 4.1 *Data Collection*

Data for this research were mainly gathered through multiple field trips during 3 years (2014–2016). We collected data both retrospectively and in real time from three main sources: (1) semi-structured interviews, (2) observation, and (3) documentation. In total, we conducted 73 interviews with 56 informants. Interviews lasted between 30 min and one and a half hours. In addition to interviews, we used observation and archival data as important triangulation and supplementary sources for understanding discrepancies among informants and gaining additional perspectives on key events and issues.

For each round of data collection, we prepared an interview guide (Patton, 2002) to make interviewing more comprehensive and systematic. We also adjusted our interview guide to adapt to different groups of informants such as state agencies, intermediary organizations, nongovernmental organizations, and social enterprises. The interviews inquired about the main organizations promoting the social enterprise form in Vietnam, their roles and activities in the social entrepreneurship field; the organizations involved in the creation of the law on social enterprise, their motivations, roles, and activities; and informants' perceptions of the law on social enterprise, the difficulties they encountered in the process of law-making, and their strategies for overcoming such obstacles.

## 4.2 Data Analysis

We used the "Gioia method" to analyze data (Gioia et al., 2013). We started coding with the interview data, identifying initial codes and grouping them into first-order categories (open coding). First, we moved quickly through interview transcripts, noting passages which refer to how, why, when and by whom the social enterprise form was promoted in Vietnam and wrote general comments about the perspective of the informants. Then, we read the transcripts again and engaged in "line-by-line coding." Whenever line-by-line coding was impossible, we conducted segment-by-segment coding. We conceptually coded each sentence or segment of data by using an "in vivo" code, which was the term repeatedly used by the informants or a simple descriptive phrase when an in vivo code was not available. Our goal is to create codes that best fit the data we have (Glaser and Strauss, 1967; Strauss and Corbin, 1998). In addition, since we aimed to uncover the types of institutional work, we looked closely at actions and coded data as actions. So, in this stage of conceptual coding, there were 118 codes that emerged from our data. Following Gioia et al., (2013), we started seeking similarities and differences among the many codes to group them into first-order categories. By the end of open coding, 23 relevant first-order categories were identified. We gave those categories labels or phrasal descriptors (preferably retaining informant terms). Then, we performed axial coding, or searching for relationships between and among first-order categories that would allow us to collapse them into a smaller number of second-order themes. During this process of axial coding (or theme development), we repeatedly reviewed the coded transcripts to ensure that all codes accurately reflected the developed themes. Eventually, twenty three first-order categories were clustered into six second-order themes: *funding experimentation*, *constructing networks*, *building capacities*, *funding policy making*, *brokering relationships*, and *providing information.* Once we identified the themes in the data, we investigated whether it was possible to distill the emergent second-order themes further into broader aggregate dimensions. Thus, in aggregating our second-order themes at a more abstract level, we ultimately identified two aggregate dimensions of institutional work for promoting social enterprise—*materializing* and *resourcing*.

# 5 Key Findings

Based on our data analysis, we identified six types of institutional work, which were undertaken by NGOs in the process of social enterprise development: (1) *funding experimentation, (2) constructing networks, (3) building capacities, (4) funding policy making, (5) brokering relationships, and (6) providing evidence*. These activities belong to two broad categories of *materializing* and *resourcing* that underpin the development of social enterprise in Vietnam. This section presents in detail these types of institutional work.

During the first period of social enterprise development (2009–2014), two non-governmental organizations BC and CSIP tried to introduce the concept of social enterprise into Vietnam by *materializing* this concept. By materializing, we refer to all the efforts made by CSIP and BC to transform potential social entrepreneurship ideas into real social enterprise models. To do this, they were involved in (1) *funding experimentation* to nurture potential local social entrepreneurship ideas and turn them into concrete social enterprise models, (2) *constructing networks* to form a group of local social enterprises and facilitate interactions within this group, and (3) *building capacities* for social enterprise as well as intermediary organizations.

## 5.1 Funding Experimentation

NGO funded pilot social enterprise models through a core program—the Social Entrepreneurs Support Program. The first program of this kind was launched by CSIP in 2009, partly in response to a lack of capital for individuals who want to start up models that looked similar to social enterprises (CSIP, 2008). The program, which took the format of an annual award for social entrepreneurs, was later renamed Social Enterprise Support Program (SESP). The objectives of this program were to identify champions with promising social entrepreneurship ideas and provide them direct support. Moreover, because the concept of social enterprise was still new to many people, including those working in the social sector, funding experimentation was an important way by which CSIP could attract potential adopters of the social enterprise form.

In the context of Vietnam, funding seed capital was important because by doing this, CSIP enabled interested individuals to mitigate risks and costs related to the adoption of the social enterprise form and thereby motivated them to take action. In the early years of social enterprise development, CSIP attracted from 100 to 200 applications to the SESP each year, of which 10 to 15 social entrepreneurship ideas were selected and nurtured. While CSIP worked directly with social enterprises, BC was rather involved in the field by funding social enterprises through CSIP.

## *5.2 Constructing Networks*

The second type of institutional work performed is *constructing networks*. This type of institutional work refers to numerous efforts of two NGOs (mainly CSIP) to build a social enterprise network and create networking opportunities. Not only does it facilitate the adoption of the social enterprise form by interested actors, but it also provides evidence of the viability of this organizational form in the eyes of stakeholders such as the public, policymakers, and international donors.

Since 2009, CSIP has emphasized the creation of a social enterprise network. CSIP has built a network of social enterprises in different ways such as training courses and networking ceremonies in the SESP program. This network has enabled social entrepreneurs to exchange resources, share their experience, and reinforce their identity. It has also contributed to diffusing the social enterprise model at field level.

In order to strengthen and expand the network, CSIP frequently organized online and offline networking activities. Online networking was mainly done via Facebook, building on the list of social enterprises supported by CSIP every year. Offline networking took diverse forms, including both formal and informal events.

While training courses and conferences were relatively formal, most networking events (e.g., coffee and networking, alumni gatherings, excursions and visits to social enterprises, social enterprise talks, and CSIP's celebrations) were organized in a very friendly and comfortable environment. Via these networking opportunities, CSIP helped social entrepreneurs to know each other, encourage each other, share best practices, as well as expand their networks of partners.

Building a network is not enough, maintaining it is also important. To maintain connections among social enterprises, CSIP initiated the formation of three social entrepreneurs' clubs in big cities including Hanoi, Da Nang, and Ho Chi Minh City. In the beginning, these clubs were coordinated by CSIP, but later they were led by groups of social entrepreneurs.

From 2009 to 2011 only, through the SESP program, CSIP succeeded in constructing a network of 42 social entrepreneurs of 29 social enterprises 4. The success in creating the first network of social enterprises over the years helped CSIP to build its own reputation and acquire knowledge on social enterprises in Vietnam. Because of its work, CSIP has accumulated a valuable database of social enterprises in Vietnam, which then enabled the center to engage in further networking.

### 5.3 Building Capacities

*Building capacities* was another type of institutional work undertaken by BC and CSIP. This type of institutional work involves not only organizing training courses for social entrepreneurs and ToT training (training-of-trainers) but also business coaching. The objective was to introduce social enterprise models to develop capacities of intermediary organizations and social enterprises.

Given the lack of knowledge on what a social enterprise is and how to start a social enterprise, capacity building is particularly crucial for BC and CSIP to disseminate the social enterprise model in Vietnam. Indeed, this activity is at the core of the SESP program. For example, as an integral part of the SESP, CSIP organized many training courses each year. The training courses generally took the form of workshops and webinars. A variety of topics were covered in these training courses: ranging from basic knowledge of social entrepreneurship (e.g., definition of a social enterprise, social issues identification, business opportunities identification, operational models, and scaling options) to more business-focused knowledge (e.g., business planning, business models, HR management, finance management, and marketing).

Parallel with training, knowledge about how to start and run the social enterprise model was also transferred through business coaching. Business coaching here involves "sending consultants" (CSIP staff or external consultants) to work directly with social enterprises in several months to help them make business plans, build or revise social business models and, in many cases, address their current management and operational issues.

Apparently, institutional work must be done through educating, which consists of all efforts to provide actors with skills and knowledge necessary to support the new institution (Lawrence and Suddaby, 2006). Organizing training courses and providing business coaching represent attempts of educating since through these efforts, CSIP seeks to diffuse the concept of social enterprise and change the mindset as well as behaviors of local adopters.

While CSIP itself organized training courses for social enterprises to improve their business capacities, BC supported CSIP to organize ToT (training-of-trainers) courses. For example, drawing on the United Kingdom's expertise on social enterprise development and its collaboration with Social Enterprise London (i.e., the first hub of social enterprises in the United Kingdom), BC helped CSIP to organize the first training program for trainers (external consultants and staff of CSIP) in Vietnam. Through its global network, BC also connected CSIP and social enterprises to participate in training workshops on social entrepreneurship in Southeast Asia.

As most Vietnamese social enterprises are small-sized, even microenterprises, and have very limited business capacity; there is a high need for technical support. For many social entrepreneurs, CSIP's technical support was valuable, helping them clearly define their social entrepreneurship orientation. It was also a vital way by which BC and CSIP implant the social enterprise concept and model into Vietnam.

Furthermore, building capacities is essential to ensuring the survival and development of newly established social enterprises.

In the next phase of social enterprise development (2012–2014), BC and CSIP engaged in *resourcing* to advocate a policy for social enterprises in Vietnam. This was done since the potential and viability of social enterprise has been confirmed through a number of successful models. Resourcing involves the deployment of different resources (e.g., finance, relationships, and information) to facilitate the development of a policy for social enterprise. Resourcing consists of three types of institutional work: *funding policymaking, brokering relationships*, and *providing information.*

## 5.4 Funding Policymaking

BC provided certain funding to policy makers during the process of making laws on social enterprise. This was done through the partnership between BC and CIEM—a think tank under the Ministry of Planning and Investment which is in charge of making the Law on Enterprises in Vietnam. For example, BC funded CIEM to carry out the first research on social enterprise in Vietnam, which served as the basis for policy advocacy. BC also supported CIEM in organizing stakeholder consultation workshops.

## 5.5 Brokering Relationships

An important type of institutional work is *brokering relationships.* By brokering relationships, we refer to all efforts of CSIP and BC to connect previously disconnected actors in the field and thereby enable the creation of the law. Our data show that this form of institutional work is comprised of three activities: connecting intermediary organizations with policymakers, connecting policymakers with relevant organizations in other countries, and connecting policymakers with social enterprises.

To support CIEM in studying social enterprise and making a new regulation on this organizational model, BC facilitated an important connection between CIEM and intermediary organizations (CSIP and Spark). This connection was crucial to the conception of policies on social enterprise since the policies should be evidence-based (i.e., based on the reality and needs of social enterprises). First of all, in 2012, BC connected CSIP with CIEM by the first research on social enterprise in Vietnam. Since CSIP has built a network of Vietnamese social enterprises and conducted several baseline studies to map out social enterprises in the first phase, it could provide CIEM useful inputs for the joint research. Therefore, BC invited CSIP to participate in the research. In the framework of this joint research, BC not only connected CSIP with CIEM but also acted as a coordinator, urging two other parties (i.e., CSIP and CIEM) to perform their work.

In addition to facilitating connections among local actors (i.e., policymakers and intermediaries), BC also brokered relationships between Vietnamese policymakers and their counterparts in other countries. BC did that mainly by organizing study tours for Vietnamese policymakers to the United Kingdom and neighboring countries. For example, in 2012, to carry out the first research on social enterprise in Vietnam, BC invited a group of Vietnamese policymakers (mainly the research team of the enterprise law project) to go to Thailand, Singapore, and Indonesia where social enterprise is much more developed and has strong government support.

After study tours to do the first research on social enterprise, in 2013, BC organized an important visit to the United Kingdom for a high-profile delegation of Vietnamese government (Office of the National Assembly and Ministry of Planning and Investment) so that they understood how useful social enterprise would be to the economy, and what legislation and financial mechanisms would be needed for its development. In this study visit, Vietnamese policymakers met UK leaders (e.g., Former Minister for Civil Society Nick Hurd, Member of Parliament Chris White) and visited Social Enterprise UK, the Skoll Centre for Social Entrepreneurship, and UK social enterprises.

Furthermore, in mid-October 2014 (i.e., just one month before the approval of the Enterprise Law), BC and CIEM led a delegation of 15 people representing the Office of Government, the Office of the National Assembly, the Ministry of Planning and Investment, Hanoi Radio—Television and Vietnam Television to attend the World Social Enterprise Forum in Seoul, South Korea. The objective of this trip was to help Vietnamese policymakers enhance their understanding of social enterprise, recognize its role in sustainable development, and learn more about the development of social enterprise around the world and the accumulated experience about social enterprise policies and laws in other countries.

In order to enable the creation of a social enterprise policy, inviting social enterprise experts from the United Kingdom to Vietnam was another way by which BC connected Vietnamese policymakers with relevant organizations in other countries. BC invited experts from the United Kingdom to present in different conferences in Vietnam and exchange with Vietnamese policymakers rather than connected policymaking organizations on a one-to-one basis. We contend that the impact of this brokerage practice might be larger since the concept of social enterprise as well as social enterprise policy models could be introduced to as many people as possible. By attending conferences, which were organized in Vietnam, and discussing with international experts, local policymakers could learn more about social enterprise models in the world and international experiences in developing policies for the social enterprise form.

Organizing study tours and providing experts aim at sharing international experience with Vietnamese policy makers (e.g., Ministry of Planning and Investment, Government Office, the National Assembly) so that they could understand how to introduce the concept of social enterprise into Vietnam's Law on Enterprises 2014.

International experience also served as important evidence for the CIEM to gain support from different public bodies for the legalization of social enterprise.

While BC facilitated the exchanges between CSIP and CIEM, CSIP played a key role in connecting social enterprises with policy makers. Brokering relationships was essential for making a policy on social enterprises because these enterprises are the final beneficiaries of such a policy. CSIP first connected CIEM to social enterprises by introducing social enterprises in its network for CIEM's field studies. Then, during the law-making process, CSIP supported CIEM in inviting social enterprises to participate in surveys and consultation workshops. After CIEM made a law draft, CSIP also called for social enterprises' comments on the draft.

### 5.6 *Providing Information*

Last but not least, two broker organizations played a key role in supporting policymakers (mainly CIEM) necessary information in order to shape a legal framework for social enterprise. Given the lack of understanding of and knowledge about social enterprise, information became a particularly important resource in the process of policy making. While BC shared social enterprise development experience of different countries, CSIP mainly provided policymakers information on social enterprises in Vietnam.

Leveraging its global network and the UK expertise in social enterprise development, BC was active in sharing international experience with Vietnamese policymakers. The experience included not only information about social enterprise models but also policies for social enterprise in different countries. In parallel with BC's efforts, CSIP not only informed CIEM about the situation of social enterprises in Vietnam but also provided the research institute with social enterprise case studies.

## 6 Discussion

Our objective in this study was to understand how local and international organizations promote the development of a new organizational model for sustainability in Vietnam. Through our analysis of an in-depth case study of two pioneer organizations in the social entrepreneurship field, we found that these organizations promote the development of social enterprises by six distinct types of institutional work (*funding experimentation, constructing networks, building capacities, funding policymaking, brokering relationships*, and *providing information*). A second finding is that the types of institutional work performed depend on the context where social enterprises operate. A third finding is that social position and networks play a critical role in the development of social enterprises. In this section, we discuss these findings in more detail.

## 6.1 Institutional Work for Social Enterprise Development

Prior research on social entrepreneurship has focused on how actors use discourse to legitimize social enterprise (Hervieux et al., 2010; Nicholls, 2010). In this study, we concentrate, however, on the types of institutional work to diffuse and develop this model in a developing country. Our analysis demonstrates that support social enterprise organizations adopted both bottom-up and top-down approaches to the development of social enterprise, engaging in different types of institutional work. Specifically, they performed six types of institutional work: *funding experimentation, constructing networks, building capacities, funding policymaking*, *brokering relationships*, and *providing information*. These types of institutional work can be classified into two broad categories or practices: *materializing* and *resourcing*. Each practice has a different objective in the process of social enterprise development. While materializing aimed at creating the new organizational model (social enterprise), the goal of resourcing was to maintain (and develop) the model. While the targets of materializing were entrepreneurs and social activists who might adopt the social enterprise model, resourcing was directed toward policymakers to assist them in the creation of a regulation on social enterprise. Materializing was the first and most important step in the process of social enterprise development, laying the groundwork for resourcing. During 3 years (2009–2011), CSIP materialized social enterprise by providing support such as funding, networks, and capacity building. As a result of its work, the incubator was able to form the first network of social enterprises in Vietnam. In the second phase of social enterprise development, CSIP made use of this network to broker (or mediate) relationships between local social enterprises and policymakers as well as provide the latter with social enterprise information, which facilitated the legalization of social enterprise. So, materializing and resourcing are strongly interrelated. The use of materializing and resourcing practices contributes to promoting social enterprise as a new and sustainable economic model in the context of Vietnam.

## 6.2 Institutional Work and Context

In line with previous research, our findings suggest that the use of specific types of institutional work depends on the context. In our case study, actors focused on introducing the concept by providing direct support to social entrepreneurs to realize social enterprise ideas. To do that, they performed the work of funding experimentation, constructing networks, and building capacities. Once the viability of the social enterprise model was verified, further support such as a policy for social enterprises was required. Because social enterprises were lacking legitimacy and thus encountered obstacles in their operations in Vietnam, social enterprise support organizations engaged in other types of institutional work such as funding policymaking, brokering relationships, and providing information to facilitate the creation of a new law on social enterprise. In this case, the choice of which types of institutional work to be done was made based on the local context, in which actors operate. Our findings resonate with the observation that "institutional contexts determine what sorts of actors would perform what kinds of institutional work" (Hwang and Colyvas, 2011, p. 64). Our case demonstrates that actors promote the legalization of social enterprise in Vietnam at a moment when the Law on Enterprises was under revision. As a result, they can grasp this opportunity to influence the introduction of social enterprise in the revised law. So, our finding is in line with prior research that highlighted temporal orientations of actors in institutional work (Battilana and D'Aunno, 2009).

## 6.3 *Social Position and Networks in Social Enterprise Development*

Last but not least, our study provides insights into the critical role of organizational position in a field and networks in social enterprise development. We found that key organizations that support social enterprises in Vietnam such as BC and CSIP have occupied the position of brokers or intermediaries that connect different stakeholders in the social entrepreneurship field. This position allowed BC and CSIP to perform certain types of institutional work to promote social enterprise. This is in line with the study of Battilana et al., (2009), which suggests that social position may be a significant explanatory factor for institutional work. Empirically, previous research on institutional work shows how the position of an actor in a field may influence their ability to shape institutions in the field, depicting institutional work of either central or peripheral actors (Riaz et al., 2016). While most studies have focused on institutional maintenance work of central actors, only a few have investigated how peripheral actors create new institutions such as social enterprise (Tracey et al., 2011). Although prior research has recognized the relationship between actors' social position and institutional work, little research has been done to explain how actors occupying the position of brokers engage in institutional work to diffuse new organizational models such as social enterprise (Bertels et al., 2014). Our study demonstrates that organizations perform different types of institutional work, especially constructing networks and brokering relationships to promote the social enterprise model. These findings extend the literature on institutional work and social entrepreneurship, demonstrating how brokerage work is used to develop the social entrepreneurship field. Based on the findings, we argue that not only dominant actors but also marginal actors playing the role of brokers in a field are able to affect institutions. Our research also illustrated how actors improve their social position through institutional work. Previous studies of institutional work have generally focused on the effect of actors' position on their institutional work. Little research has been done to explore how marginalized actors implement institutional work to become more central in organizational fields (Waldron et al., 2015). Our study contributes to research on institutional work by illustrating how organizations moved into the position of brokers in the social entrepreneurship field through three types of institutional work—*funding experimentation, constructing networks,* and *building capacities*, and then how they leveraged their new position to facilitate institutional work of other actors. By highlighting how institutional work is a central contributor to actors' position in a field, our study also responds to recent calls for more research on "other factors that may contribute to field position" (Riaz et al., 2016, p. 1553). We argue that actors can become more influential in organizational fields as a result of their diverse types of institutional work. One of the most important types is constructing networks, which facilitated actors' engagement in other types of institutional work (e.g., brokering relationships and providing information). In addition to social position, our findings also revealed the role of networks in the development of social enterprises in Vietnam. In traditional business in Vietnam, the role of social capital in firm success has been clearly articulated, with weak-ties and informal networks being particularly critical to performance (Santarelli and Tran, 2013). In the area of social entrepreneurship, social capital is even more important in driving social innovations and social enterprise (Dacin et al., 2011; Mair and Marti, 2006). Previous studies on social enterprises in Vietnam demonstrate the role of bonding social capital within the ecosystem and social enterprises themselves (Easter and Conway Dato-On, 2015; Truong and Barraket, 2018). However, such studies did not explore how social networks were mobilized to diffuse and develop the social enterprise model. Our study shows that social networks, especially the emergence of a social enterprise network is crucial to the development of the social entrepreneurship field in Vietnam as social enterprise support organizations drew on such a network to advocate policy for social enterprises. By showing the efforts of social enterprise support organizations to connect different stakeholders in the social entrepreneurship field, our study demonstrates how actors rely on networks to promote social enterprise development.

## 7 Conclusion

In conclusion, this chapter focuses on purposive activities to develop social enterprises in Vietnam in response to the call for more empirical research on institutional work and social entrepreneurship in the developing world. The study found that social enterprise support organizations performed six types of institutional work to promote the social enterprise model. In addition, it highlights the link between institutional work and context. More importantly, the study sheds light on the role of social position and networks in social enterprise development. By analyzing deliberate actions to develop social enterprises in a developing country, the research makes contributions to both literature on institutional work and social entrepreneurship. Not only does it contribute to our understanding of the process by which a sustainable organizational model such as social enterprise arises and achieves stability, legitimacy, and adherents but also provides practical implications for developing social enterprises elsewhere.

## References


Austin J, Stevenson H, Wei-Skillern J (2006) Social and commercial entrepreneurship: same, different, or both? Entrep Theory Pract 30(1):1–22

Bacq S, Janssen F (2011) The multiple faces of social entrepreneurship: a review of definitional issues based on geographical and thematic criteria. Entrep Reg Dev 23(5–6):373–403

Battilana J, D'Aunno T (2009) Institutional work and the paradox of embedded agency. In: Lawrence TB, Suddaby R, Leca B (eds) Institutional work: actors and agency in institutional studies of organization. Cambridge University Press, New York, pp 31–58

Battilana J, Dorado S (2010) Building sustainable hybrid organizations: the case of commercial microfinance organizations. Acad Manag J 53(6):1419–1440

Battilana J, Lee M (2014) Advancing research on hybrid organizing – insights from the study of social enterprises. Acad Manag Ann 8(1):397–441

Battilana J, Leca B, Boxenbaum E (2009) How actors change institutions: towards a theory of institutional entrepreneurship. Acad Manag Ann 3(1):65–107

Bertels S, Hoffman AJ, DeJordy R (2014) The varied work of challenger movements: identifying challenger roles in the US environmental movement. Organ Stud 35(8):1171–1210

Bornstein D (2007) How to change the world: social entrepreneurs and the power of new ideas. Oxford University Press, Oxford, UK

British Council, CSIP, & Spark (2011) Final report: Vietnam 2011 Social enterprises mapping project

British Council (2019) Social enterprise in Vietnam

Brouard F, Larivet S (2010) Essay of clarifications and definitions of the related concepts of social enterprise, social entrepreneur and social entrepreneurship. In: Fayolle A, Matlay H (eds) Handbook of research on social entrepreneurship. Edward Elgar, pp 29–56

Cochran PL (2007) The evolution of corporate social responsibility. Business Horizon 50(6):449–454

Dacin PA, Dacin MT, Matear M (2010) Social entrepreneurship: why we don't need a new theory and how we move forward from here. Acad Manag Perspect 24(3):37–57

Dacin MT, Dacin PA, Tracey P (2011) Social entrepreneurship: a critique and future directions. Organ Sci 22(5):1203–1213

Dees JG (2007) Taking social entrepreneurship seriously. Society 44(3):24–31

Di Domenico M, Haugh H, Tracey P (2010) Social Bricolage: theorizing social value creation in social enterprises. Entrep Theory Pract 34(4):681–703

Dimaggio PJ, Powell W (1983) The iron cage revisited: institutional isomorphism and collective rationality in organizational fields. Am Sociol Rev 48(2):147–160

Easter S, Conway Dato-On M (2015) Bridging ties across contexts to scale social value: the case of a Vietnamese social enterprise. J Soc Entrep 6(3):320–351

Ebrahim A, Battilana J, Mair J (2014) The governance of social enterprises: mission drift and accountability challenges in hybrid organizations. Res Organ Behav 34(2014):81–100

Florek NE (2013) Enabling social enterprise through regulatory innovation: a case study from the United Kingdom. J Sustain Financ Invest 3(2):155–175

Gioia DA, Corley KG, Hamilton AL (2013) Seeking qualitative rigor in inductive research: notes on the Gioia methodology. Organ Res Methods 16(1):15–31

Glaser BG, Strauss AL (1967) The discovery of grounded theory: strategies for qualitative research. Aldine, Chicago

Hervieux C, Gedajlovic E, Turcotte M-FB (2010) The legitimization of social entrepreneurship. J Enterpr Commun People Places Global Econ 4(1):37–67
Hwang H, Colyvas JA (2011) Problematizing actors and institutions in institutional work. J Manag Inq 20(1):62–66
Lawrence TB, Suddaby R (2006) Institutions and Institutional work. In: Clegg SR, Hardy C, Lawrence TB, Nord WR (eds) The Sage handbook of organization studies, 2nd edn. Sage, London, pp 215–254
Lawrence TB, Suddaby R, Leca B (eds) (2009) Institutional work: actors and agency in institutional studies of organizations. Cambridge University Press, New York
Mair J, Martí I (2006) Social entrepreneurship research: a source of explanation, prediction, and delight. J World Bus 41(1):36–44
Martí I, Mair J (2009) Bringing change into the lives of the poor: entrepreneurship outside traditional boundaries. In: Lawrence TB, Suddaby R, Leca B (eds) Institutional work: actors and agency in institutional studies of organization. Cambridge University Press, New York, pp 92–119
Martin RJ, Osberg S (2007) Social entrepreneurship: the case for a definition. Stanf Soc Innov Rev Spring 2007:29–39
Meyer JW, Rowan B (1977) Institutionalized organizations: formal structure as myth and ceremony. Am J Sociol 83(2):340–363
Moss TW, Short JC, Payne GT, Lumpkin GT (2011) Dual identities in social ventures: an exploratory study. Entrep Theory Pract 35(4):805–830
Nguyen DC, Luu MD, Pham KO, Tran THG (2012) Social enterprise in Vietnam: concept, context and policies. Published by British Council, CIEM, and CSIP
Nicholls A (2010) The legitimacy of social entrepreneurship: reflexive isomorphism in a pre-paradigmatic field. Entrep Theory Pract 34(4):611–633
Patton MQ (2002) Qualitative research & evaluation methods, 3rd edn. Sage, Thousand Oaks, CA
Peredo AM, McLean M (2006) Social entrepreneurship: a critical review of the concept. J World Bus 41(1):56–65
Pless NM (2012) Social entrepreneurship in theory and practice – an introduction. J Bus Ethics 111(3):317–320
Riaz S, Buchanan S, Ruebottom T (2016) Rhetoric of epistemic authority: defending field positions during the financial crisis. Hum Relat 69(7):1533–1561
Roberts D, Woods C (2005) Changing the world on a shoestring: the concept of social entrepreneurship. Univ Auckl Bus Rev 7(1):45–51
Ruebottom T (2013) The microstructures of rhetorical strategy in social entrepreneurship: building legitimacy through heroes and villains. J Bus Ventur 28(2013):98–116
Santarelli E, Tran H (2013) The interplay of human and social capital in shaping entrepreneurial performance: The case of Vietnam. Small Bus Econ 40(2):435–458
Short JC, Moss TW, Lumpkin GT (2009) Research in social entrepreneurship: past contributions and future opportunities. Strateg Entrep J 3(2):161–194
Strauss AL, Corbin JM (1998) Basics of qualitative research: techniques and procedures for developing grounded theory. Sage
Sud M, VanSandt CV, Baugous AM (2009) Social entrepreneurship: the role of institutions. J Bus Ethics 85:201–216
Thompson J, Doherty B (2006) The diverse world of social enterprise: a collection of social enterprise stories. Int J Soc Econ 33(5/6):399–410
Tracey P, Phillips N, Jarvis O (2011) Bridging institutional entrepreneurship and the creation of new organizational forms: a multilevel model. Organ Sci 22(1):60–80
Truong A, Barraket J (2018) Engaging workers in resource-poor environments: the case of social enterprise in Vietnam. Int J Hum Resour Manag 29(20):2949–2970
Waldron TL, Fisher G, Navis C (2015) Institutional entrepreneurs' social mobility in organizational fields. J Bus Ventur 30(1):131–149

World Bank (2024) Vietnam: Overview. Retrieved March, 27 2025 from https://www.worldbank.org/en/country/vietnam/overview

World Bank (2025) Vietnam: Data. Retrieved March, 27 2025 from https://data.worldbank.org/country/viet-nam?view=chart(downloadExcel)

Zahra SA, Gedajlovic E, Neubaum DO, Shulman JM (2009) A typology of social entrepreneurs: motives, search processes and ethical challenges. J Bus Ventur 24(5):519–532

Zucker L (1977) The role of institutionalization in cultural persistence. Am Sociol Rev 42(5):726–743

**Trang-Thu Tran** has been teaching at Foreign Trade University since 2007. Her research interests include corporate social responsibility, social entrepreneurship, and social innovation. She earned her PhD in Management from HEC Montréal (2017). Her doctoral dissertation explores the institutionalization of social enterprise in Vietnam. She has been one of the trainers for Vietnam Youth Social Innovation Program (VYSIP) supported by British Council and has served as an expert on several projects of Center for Social Initiatives Promotion (CSIP).

GPSR Compliance

*The European Union's (EU) General Product Safety Regulation (GPSR) is a set of rules that requires consumer products to be safe and our obligations to ensure this.*

*If you have any concerns about our products, you can contact us on ProductSafety@springernature.com*

In case Publisher is established outside the EU, the EU authorized representative is:

Springer Nature Customer Service Center GmbH
Europaplatz 3
69115 Heidelberg, Germany

**Batch number: 10703869**

Printed by Printforce, the Netherlands